4/44/14 I

(Four Years and Forty-Four Fourteeners)
First Fourteeners

David A. Lien

To Diane,
A fellow Coloradan
and friend,
David A. Lien

Outskirts Press, Inc.
Denver, Colorado

Outskirts Press, Inc.
http://www.outskirtspress.com

ISBN: 978-1-4327-6410-4

Outskirts Press and the "OP" logo are trademarks belonging to Outskirts Press, Inc.

PRINTED IN THE UNITED STATES OF AMERICA

"And yet, and yet... It is not enough to say we climb a given mountain because no one else has done it. Even if true, that is not sufficient. Nor will it do to say, simply, 'Somebody has to do it.' They won't believe you. Most mountains have now been climbed anyhow, some of them many times... Then why? We have not yet answered this elementary question, and perhaps no one ever will answer it to the satisfaction of nonclimbers. But one must attempt the answer."[1]

"Men and women climb mountains–whether in the Rockies or in the Himalayas–for the same reason they blast off in rockets to the moon, launch poems and prayers at the stars, send symphonies of thought, music, mathematics, and fiction into the highest and deepest reaches of the human soul. Because... it's something to do. Because, there's nothing better to do. Because of all our terrors, none is more terrible than boredom, the nothingness of a static existence, the infantile paralysis of Saturday night in Page, Arizona."[2]

–Edward Abbey, *The Journey Home*

---

1   Edward Abbey. *The Journey Home*. New York: Penguin Books, 1977, p. 214.
2   Edward Abbey. *The Journey Home*. New York: Penguin Books, 1977, p. 215.

Acknowledgments

The philosopher and Nobel Prize winner Albert Camus once said a man's work is nothing but a slow trek to rediscover those two or three great, simple images in whose presence his heart first opened. My formative years were spent hiking, hunting, camping, canoeing, fishing, and trapping amidst the woodlands, wildlands, lakes, rivers, and streams of northern Minnesota.

It was there where I first learned about the wild and natural (the real) world, where my heart first opened, and it's from there where I draw my strength and determination to fight on for the wildlands that remain. Without those early life experiences nurtured by direct contact with the outdoors, it's doubtful I would have written 4/44/14.

Because much of what follows has been said far more eloquently and persuasively by many other dedicated outdoorsmen and women, most of whom are far more interesting and qualified to do so than I, there are many people to thank and recognize. In many ways this is their book, and I apologize in advance to anyone I haven't appropriately acknowledged.

As Laurence J. Peter said, "Originality is the fine art of remembering what you hear but forgetting where you heard it." Edward Abbey adds, "Some of these notes… may be unconscious plagiarisms from the great and dead (never steal from the living and mediocre), ideas absorbed in my reading so long ago that I've made them mine and forgotten the source. If so, the author would appreciate hearing from readers on this point."[3]

Before mentioning some of the people who have had the most influence on my life, and hence this book, it's worth considering how a book comes about in the first place. At the most basic level, writing is the result of life, which (I think) is mostly about learning, and then applying that knowledge to improve ourselves

---

3   Edward Abbey. *A Voice Crying in the Wilderness (Vox Clamantis in Deserto)*. New York: St. Martin's Press, 1989, p. x.

and, hopefully, the lives of others. It's about learning from our misdeeds and mistakes, our accomplishments and victories, and then endeavoring to make a difference for the better. This book is another step in my life journey and ongoing quest to learn and make a difference.

I've learned much from other people over the years and have drawn heavily from their knowledge and words for this book. For their hard work, and in many cases, dedicated friendship, I am eternally grateful. A few of the people who I've been blessed to know and call family and friends during my life so far include my aunt, Lois Nyman, who single-handedly managed to raise her four boys along with my sister and me. How she pulled that off without losing her mind, I'll never know.

My grandmother, Eleanor Boyce, who was one of the wisest, kindest, gentlest, most endearing souls I've ever known. My sister, Sandy Filipiak, who has had to negotiate life with an unconventional and absentee brother, without (like me) the benefit of our parents, James and Sharon. Then there's Ed Schmidt, my fourth-grade teacher.

Ed is a co-founder and past president of the Minnesota Deer Hunters Association (MDHA) and co-editor of *Whitetales* magazine. He published my first article, "Whitetail Deer Evolution," in the spring of 1995, inspiring me to write.[4] Ed's life is a testament to the fact that the best teachers never stop teaching. Thanks, Ed. My friend David Petersen, one of the nation's foremost hunting ethicists and founder of the Colorado Backcountry Hunters and Anglers, has continually provided me with invaluable hunting and writing advice and encouragement.

And my Colorado climbing companion (and good friend) Lori Shepard and fellow hunter (and coworker) Peter Jensen were both kind enough to read 4/44/14 drafts. Their feedback was priceless and smoothed out many of the rough edges herein. I would be remiss not to mention my fellow Minnesotan–Air Force officer and beer-drinking connoisseur–Eric Johnson, who once asked me what I do with myself on Sundays. You're reading it buddy.

Last, but not by a long shot least, is my niece, and deer hunting extraordinaire, Faith Filipiak, representing the next generation.[5] "Dream big and dare to fail," and you're certain to reach great heights. Thank you, one and all, and all those who I haven't mentioned. You're no less important, and I'm humbled and privileged to call you all family, friends, and mentors.

---

4   See article: David A. Lien. "Whitetail Deer Evolution." *Whitetales*: Spring 1995, p. 39.

5   See: Grand Rapids *Herald-Review*. "A successful season." 11/28/07, p. 4b & David A. Lien. "First Hunt Hat Trick." *Whitetales*: Spring 2008, p. 46.

"We will stomp to the top with the
wind in our teeth."
–George L. Mallory, 1924

"Well George, we knocked the bastard off."
–Sir Edmund Hillary, 1953

"When in doubt, go higher."
–Mountain Gazette, 2002

Foreword

"Measure the quality of your life by the impact
you have on the lives of those around you."
–Mike Miller[6]

In her article "Answering The Call," M editor Martha Conventry says the urge to make a difference can come at any age. Not just at 50, when you wake up in the middle of the night afraid you've squandered your days, but also at 25, when the 9-to-5 life of your parents looks like a prison.[7] It is! Whenever the urge hits, whatever you do, *make a difference* that counts for something, and make it something you love and are passionate about.

People often talk about being deeply moved by making a difference in just one person's life. That's a good thing and a great feeling, and something many more of us should strive to achieve. However, I believe each of us can make a much larger and more significant impact, one that will affect both current and future generations, by helping to improve the life of this spinning ball of blue we all call home. While you may be thinking this is starting to sound like some stereotypical environmental spiel, people often forget that hunters were the first environmentalists.

Mark Johnson (executive director of the Minnesota Deer Hunters Association) says, "Environmentalism is one very important key for MDHA. As hunters, trappers, and fisherpersons, we are the original environmentalists. We support it with our license dollars, we declare it with our time spent afield, and we teach it to our children."[8] Teddy Roosevelt was a hunter-conservationist too, and said, "The true... hunter must be a lover of nature as well as of sport, or else he will

6   Mike Miller. "Measure success by your own standards." *Credit Union* magazine: July 2001, p. 25.
7   Martha Coventry. "Answering the Call." *M*: Fall 2004, p. 1.
8   Mark Johnson, MDHA executive director. "The Outlook." *Whitetales*: Summer 2006, p. 3.

miss half the pleasure of being in the woods."[9]

Conservationist Michael Frome adds, "Hunting at its finest involves environmental satisfactions: an authentic immersion in nature, with hard efforts in tracking, scenting, and pinpointing the game for hours or even days when necessary. The quality of the hunt is scantly measured by the size of the bag but rather by the experience of woodmanship, living with the animals and thinking like them, perceiving surroundings from the viewpoint of the animal. The hunter-conservationist imposes limitations on himself... he strives to preserve a world safe for leopards and bison, a world for rocks and all nature."[10]

"It's essential to remember," explains my expert bow and elk hunting friend David Petersen (in his book *A Man Made of Elk*), "that we can hunt in America today only because wildlife is so abundant. And wildlife is so abundant today only because of the then-unpopular actions of an enlightened handful of yesterday's hunters—specifically, Theodore Roosevelt and a hundred or so fellow sportsmen/conservationists who had the moral will and political weight to introduce protective wildlife legislation and get it enacted and enforced."[11]

In "Camo Conservation," *Trout* magazine contributor Kirk Deeter notes that when it comes to protecting the nation's environmental resources, the voice of the person who wades a river with a fly rod, or walks the timber with a bow or rifle, matters now more than ever, just as Theodore Roosevelt and the other sportsmen who helped to found the modern conservation movement envisioned. And as was the case 100 years ago, protecting the nation's natural resources ultimately is not a Republican or Democratic issue. It's cultural, and uniquely American.[12]

"Theodore Roosevelt understood the vital role that hunting plays in American life, as well as the importance of protecting lands and wildlife to sustain that tradition," Interior Secretary Ken Salazar observed. "The early efforts of America's hunters and anglers to preserve our nation's wildlife heritage fueled the modern conservation movement and left us the natural bounty we are now entrusted with protecting."[13]

By striving to protect and preserve our public (and other) lands, we'll make terra firma a more livable place for "leopards and bison," for current and future generations of humans and other species, and we'll be making a difference in its life. When we do that, we don't just make a difference in the life of one individual, but instead, touch the lives of billions of people and all the other life-forms that inhabit the planet, present and future.

9   See article: David A. Lien. "Teddy Roosevelt: Hunting's Rough Rider." *Whitetales*: Spring 2004, p. 32.

10   Michael Frome. *Battle for the Wilderness*. Salt Lake City, UT: University of Utah Press, 1974, p. 74.

11   David Petersen. *A Man Made of Elk*. Eagle, Idaho: TBM, Inc., 2007, p. 160.

12   Kirk Deeter. "Camo Conservation." *Trout*: Spring 2009, p. 26.

13   Sam Cook. "Hunting council will advise Interior Department." *Duluth News Tribune*: 2/9/10.

The late Senator Paul Wellstone (from my boyhood home of Minnesota) said the future will not belong to those who stand on the sidelines. "The future," he emphasized, "will belong to those who are willing to work hard for what they believe in." One simple way for each of us to follow Paul's sage advice is by letting our elected legislators know we want more of our tax dollars used for the preservation of wildlands, wildlife, and clean air and water across the country and the world.

Why worry about clean air, water, and wildlife and wilderness? Because, as Pulitzer Prize-winning writer Wallace Stegner said, "We were born of wilderness, and we respond to it more than we sometimes realize. We depend upon it increasingly for relief from the termite life we have created. Factories, power plants, resorts, we can make anywhere. Wilderness, once we have given it up, is beyond our reconstruction." Wilderness "is the very stuff America is made of," Aldo Leopold wrote in 1925, and as it's degraded, denuded, and developed, all that sprang from it will feel the loss and face the repercussions.

The future we're all hoping for hinges on taking action and telling our legislators and other decision makers we want more roadless areas, wilderness areas, national parks, national monuments, and national wildlife refuges set aside and protected, more wild places and wide-open spaces preserved for current and future generations. Let 'em know we want *them* to make a difference in the life of the planet. Just by doing that, the average citizen can make a difference.

> "If you're a senator and the mailbag [or inbox] contains
> 10,000 letters from constituents on a single topic, you
> know you've got a hot issue."
> –John H. Adams[14]

If a lawmaker receives even 50 e-mail messages, calls, or faxes a day on a single issue, it makes a big impression. Politicians know that for every person contacting them on a particular topic, there are many more who feel the same way but who haven't taken the extra step to write, e-mail, or call. Regardless, all these people are likely voters, which legislators recognize. Remember, only a matter of several hundred votes put President George W. Bush into office the first time.[15] Every vote does count!

Lawmakers often calculate that each letter or e-mail they receive represents

---

14   John H. Adams, Natural Resources Defense Council (NRDC) president. "Playing to Win." *Onearth*: Fall 2004, p. 4.

15   In 2000, George W. Bush lost the popular vote after leading in the polls for months, ultimately winning the electoral vote because of a contested 537-vote margin in Florida. A subsequent recount showed him losing Florida. In 2004, he won reelection with 51.3 percent of the popular vote—the lowest percent of any victorious Republican incumbent in American history.

the point of view of at least 100 other constituents. Consequently, a single e-mail from one individual counts more than we may think. MDHA executive director Mark Johnson knows this is an issue that's beyond us individually, "It's about our quality of life, about what we want our country to look like not only 20 years from now, but 100 years from now."[16]

Mark says, "Our grassroots gives us an unmistakable, yet virtually unused power: the power of the individual. Most people have no idea how much influence one call or letter from you has on your legislator. It demands consideration... the constituent wields the biggest stick... Without you it doesn't happen. Get involved!"[17] Mark gets it, and knows what we do as individuals matters greatly.

"Our individual contributions, both bad and good, add up with the contributions of others," he adds. "If we abuse our environment, or refuse to dig for the answers to save a dwindling species, or decide not to get involved as volunteers or constituents, we are contributing to the end results. Our actions and our inactions count greatly. What we do with what we have, as well as what we decide we do not need, will determine what our children and grandchildren will have. It starts with us. It starts now."[18]

There are countless instances where the efforts of just one person changed the course of history for the better: Rachel Carson, Margaret Murie, John Muir, Sigurd Olson, Edward Abbey, Aldo Leopold, Bob Marshall, Martin Luther King, Jr., Rosa Parks, and untold numbers of others less recognized but no less important. Imagine what hundreds of thousands or millions of us can do working together collectively for the preservation of wildlands and wildlife. Sound insurmountable? Too big a task?

Pulitzer Prize winner Alice Walker (author of *The Color Purple*) once said the most common way people give up their power is by thinking they don't have any. Social activist Dorothy Day seemed to agree when she said, "People say, what is the sense of our small effort. They cannot see that we must lay one brick at a time." There is only one earth, but there are billions of people. So when each person does something, anything to help make a little difference, it adds up to a very big difference, and with the advent of the Internet and the Web sites below, we now have the means to coordinate and focus our efforts. *Collective efforts create exponential results.*

To start making a difference, visit one or all of the following sites and *Take Action:* The Wilderness Society–*www.wilderness.org;* The National Parks Conservation Association–*www.npca.org;* The Natural Resources Defense Council–*www.nrdc.org;* The Trust for Public Land–*www.tpl.org;* Trout Unlimited–

16   Babe Winkelman. "Spring!" *Whitetales:* Spring 2004, p. 8.

17   Mark Johnson, MDHA executive director. "The Outlook." *Whitetales:* Summer 2001, p. 3.

18   Mark Johnson, MDHA executive director. "Our Actions and Inactions." *Whitetales:* Fall 2007, p. 3.

www.tu.org; The Nature Conservancy–*www.tnc.org*; The Theodore Roosevelt Conservation Partnership–*www.trcp.org*; Backcountry Hunters and Anglers–*www. backcountryhunters.org*; REP America–*www.rep.org*; The League of Conservation Voters–*www.lcv.org*.

Most have *Take Action* centers that allow individuals to send e-mail letters to their elected legislators and other decision makers with one click of a mouse. The centers oftentimes are able to "recognize" each visitor, automatically select his or her senator and representative, and sign your names to advocacy letters that are already written for you. Doing this takes at most a matter of minutes and puts your elected officials on notice that you care about the issues impacting wildlands and wildlife and the future of our country and planet.

> "Never doubt that a small group of thoughtful,
> committed citizens can change the world;
> indeed, it's the only thing that ever has."
> –Margaret Mead

> "We must not, in trying to think about how we can make
> a big difference, ignore the small daily differences
> we can make which, over time, add up to big
> differences that we often cannot foresee."
> –Marian Wright Edelman

Think about someone you know or have heard about who has helped build a school playground or contributed to saving land for a city park. Someone who lobbied local legislators for the designation of a new national park or wilderness area. Someone who labored to save the Grand Canyon from being dammed, Yosemite's Giant Sequoias from being logged, or Yellowstone's grizzlies from being shot, trapped, and poisoned into extinction.

Think about anyone who wrote a letter to help get some of the miles of mountainous forests of Colorado and Montana set aside as Rocky Mountain and Glacier National Parks. Anyone who called their congressman or -woman to help get part of the redrock canyons of southern Utah protected in Arches and Canyonlands National Parks or the Northwoods of Minnesota preserved in the Boundary Waters Canoe Area Wilderness and Voyageurs National Park.

Think about everyone who has helped save roadless areas from industrial logging, watershed-poisoning sulfide mines, oil and gas drilling, and the on-slaught of ATV overuse and abuse that is increasingly turning our wildlands into wastelands, trashing the democratic birthright that belongs to all Americans.

These public lands defenders will generally be people you admire and are proud to know and tell your children about, and today, we all have the opportunity to become one of them.

But why should we bother? Because, it seems to assume that if the wild and natural world that gave life to our (and all the other) species on this planet is disappearing, eventually those that sprang from it will disappear too. Consider how many endangered and extinct species there are already, and the new and renewed epidemics of infectious diseases, deepening droughts, and extreme weather patterns sweeping across the globe. Temporary climatic variation, natural planetary evolution, warning signs?

Can we afford to wait for definitive answers? Can we afford to be wrong about something like global warming? Shouldn't we take action now, before it's too late? Why can't we give up a little, sacrifice some today, if it's done to ensure future generations will have a habitable planet? Republican or Democrat, conservative or liberal, I vote for those who are looking out not just for the short-term needs of this generation, but for the countless generations yet to come.

By now, you may be thinking, *So what the heck does all this have to do with climbing mountains, with* Four Years and Forty-Four Fourteeners? Climbing a mountain is nothing more than a series of small efforts, slow and steady steps, strung together over time to reach a higher goal: getting to the top. Mountain climbing is a metaphor for life, an example of how we can all make a difference: take small, steady steps, always moving upward and onward.

Our individual actions, our steps, may seem insignificant in the whole scheme of things, but when added together over time and combined with the steps and actions of everyone else, they can result in overcoming long odds and reaching incredible heights. Earth First! cofounder Dave Foreman says, "Talk is cheap. Action is dear." Similarly, Edward Abbey said, "Sentiment without action is the ruin of the soul," and both Dave and Ed learned from experience that "writing, reading, [and] thinking are of value only when combined with effective action."[19]

In a July 2008 *USA Today* letter-to-the-editor ("Citizens must protect democracy"), Arnold Stieber said, "Democracy is much more than voting. Democracy requires action. Like our Founding Fathers, Americans must actively speak out and defend the rights that have been bestowed upon us by our creator. We are not pawns and slaves. We are a free and independent people, and we must work to remain so. To do otherwise is to invite domination and the abuse of power."[20]

---

19   Ken Wright. "Postcards from Ed." *Inside/Outside Southwest*: August/September 2006.
20   Arnold Stieber. "Citizens must protect democracy." *USA Today*: 7/3/08, p. 10A.

What Dave, Ed, and Arnold are saying, besides get up off of our butts and do something, is *all* our actions matter, and each of us must take action for anything good to happen. In 1983, David Petersen landed assignments from *Mother Earth News* and *Writer's Digest* to interview Edward Abbey. Somewhere in the midst of a rambling conversation Petersen asked, "Are you an atheist?" Abbey rumbled, "I am an *earthiest*. I stand for what I stand on."[21] So do I.

And so does (I'm guessing) Mark Johnson. "I would venture to say many deer hunters are 'environmentalists,'" Mark says. "Some may resist that title, but if we are concerned about the environment and we actually act on its behalf, then 'environmentalists' we are. Likewise, if we understand the need to conserve our natural resources so we do not use them faster than they are replaced, and we participate toward that conservation, we are, indeed, conservationists. Now the $64 question: Are you?"[22]

The founder of modern conservatism, Edmund Burke, wrote: "Never, no never, did nature say one thing and wisdom say another." Similarly, David Jenkins (REP America's government affairs director) says, "Whether it was in Sunday school hearing about Jesus telling his disciples not to waste food, learning from a father or grandfather to only hunt and kill what you need, understanding the financial importance of saving for the future, or being inspired by the land conservation ethic of Theodore Roosevelt, most rank-and-file Americans understand that a conservation ethic is an important aspect of being American."[23]

Indeed, and all of us have a responsibility to take action to promote and perpetuate that conservation ethic. Chaos theory teaches that an influence as tiny as the flapping of a butterfly wing can set in motion a chain of events that will eventually alter the course of global weather patterns. So it goes for the actions each of us can take on behalf of protecting and preserving wildlands and wildlife: e-mail letters, faxes, or phone calls to your elected legislators and other decision makers will result in setting in motion a chain of events that can lead to the preservation of wild places and wide-open spaces for current and future generations.

I'm confident in our collective ability to do what needs to be done because over the years, I've discovered that most of us are more than we seem, more than what's on the surface, on the cover of the ephemeral books that are our lives. Some people are sentences or paragraphs, some are essays, others short stories or novels, but we are all more than the sum of our parts, more than our individual words and actions.

When our efforts are combined with those of others fighting for the same

21   David Petersen. *Elkheart*. Boulder, CO: Johnson Books, 1998, p. xi.
22   Mark Johnson, MDHA executive director. "Defined By Our Actions." *Whitetales*: Spring 2009, p. 3.
23   David Jenkins. "Taking Control of Our Energy Future." *C.E.P. Quarterly*: Spring 2006.

cause, we can contribute to the writing of books and moving of mountains that would not have been possible otherwise. The wild and natural world must be saved, both by us and from us. Mountains are the metaphor: we climb them one step at a time, each step building on the other, bringing us closer to our goal. Acting together, we can conquer the seemingly impossible obstacles that must be overcome if the wild Earth is to be saved for our children's children and beyond.

We should all endeavor to live our lives fully and to tell good stories by how we live them, stories that our children and theirs will want to remember, relive, and pass down to future generations. Start your story today. Part of mine is in the pages that follow. I hope you enjoy it and endeavor to make a difference in the life of this wild and wondrous country and planet.

Best wishes,

David A. Lien

"At last you are on the peak and your first impulse is to look down. But no. Look up! You have read and heard many times of the 'deep blue sky'... It is seen only from highpoints–from just such a place as you are now standing upon. Therefore look up first of all and see a blue sky that is turning into violet. Were you ten thousand feet higher in the air you would see it darkened to a purple-violet with the stars even at midday shining through it."
–John C. Van Dyke, *The Desert*[24]

"The tops of mountains are among the unfinished parts of the globe, whither it is a slight insult to the gods to climb and pry into their secrets, and try their effect on our humanity. Only daring and insolent men, perchance, go there."
–Henry David Thoreau, "Ktaadn"

---

24   John C. Van Dyke. *The Desert*. New York: Scribners, 1901, p. 225.

Contents

(To be continued in 4/44/14 II: Nemesis)

And So It Begins

"This trip will be longer than I expected, for I will be in many
beautiful places, and do not wish to taste, but to drink deep."
                                                              —Everett Ruess

After graduating from the University of Minnesota-Duluth (UMD) in the summer
of 1990, during May 1991, I reported for Undergraduate Missile Training (UMT)
at Vandenberg Air Force Base in California, stopping to visit Yosemite National
Park on the cross-country drive from Minnesota to Vandenberg. While there,
I roamed through the fog-shrouded Yosemite Valley and snow-covered Giant
Sequoia forest of Mariposa Grove.

Words truly cannot describe the unbelievable breadth and unimaginable beauty of the Yosemite region, nor the grandeur of her Giant Sequoias; silent sentinels,
standing ramrod-straight and neck-stretching tall, weathering sun, rain, snow, and
storms. Seemingly immovable goliaths, these ancient wooden monoliths of the
Sierra Nevada are some of the largest and oldest living things on earth.

Imagine 150 years ago, chasing a wounded grizzly bear through these *Alice
in Wonderland*-like forests and suddenly encountering the shaggy, cinnamon-
red trunk of a tree so massive that only with outstretched arms could 16 people
encircle it. During 1852, in an area now protected as Calaveras Big Trees
State Park, A. T. Dowd—a frontiersman and backwoods hunter from Connecticut—
abandoned pursuit of a grizzly and gazed upward at a massive tree that stretched
more than 300 feet skyward.[25]

Some of these Sequoias are 3,500 years old with diameters greater than 35
feet and heights surpassing that of the Statue of Liberty, taller than a 25-story
building. Like Lady Liberty and the Minuteman III ICBMs I would soon be trained

25   Save-the-Redwoods League. "The Big Trees." *Save-the-Redwoods League*: Fall Bulletin, 2002, p. 4.

to launch in the event of a nuclear conflict, these trees were reminders to me of what was at stake in the fight for America's democratic birthright—our remaining wild places and wide-open spaces.[26]

I knew then more clearly than ever what it was I had devoted the next four years of my life to defend. As Kim Heacox says in *Visions of Wild America*, our tradition of democratic self-government, originally established in England under the weight of hundreds of years of human history and monarchy, was established in America by the force of hundreds of miles. Hundreds of miles of larger-than-life landscapes; horizon-to-horizon wilderness, and natural majesty equaled nowhere else in the world.[27]

General Francis Marion, the Revolutionary War's legendary Swamp Fox, wrote in his wartime journals: "I look at the venerable trees around me and I know that I must not dishonor them." History shows that General Marion defended and honored the trees and wilderness that sheltered his military forays for personal and political freedom. Following in his footsteps, I paid tribute to the cinnamon-red sentinels of Yosemite on my cross-country journey to begin defending those same freedoms.

I knew then, without a doubt, that defending America's freedom and democracy was synonymous with defending the backdrop against which they formed; the wild places and wide-open spaces of this vast continent and great nation. Before leaving Mariposa Grove, I stopped and took one last look around. Standing in silent, awed reverence, I felt "the blinding light of human obligation, as if the crosshairs of destiny has settled on my shoulders."[28]

After reporting to Vandenberg and completing Undergraduate Missile Training, in September, I headed for the Midwest and my permanent duty station at Grand Forks Air Force Base (GFAFB), North Dakota, but not without stopping off to see the Grand Canyon for the first time. Approaching from the south, the terrain looked mostly flat and nondescript. It didn't seem like anything very remarkable could lie just ahead, below the horizon, but when the Canyon came into view, it was a jaw-dropper.

"I was halfway across America,
at the dividing line between the East of
my youth and the West of my future."
—Jack Kerouac, *On the Road*[29]

---

26   David A. Lien. "Silent Sentinels." *Association of Air Force Missileers (AAFM) Newsletter*: March 2009, p. 1 (article continues on p. 5, not p. 2); http://www.afmissileers.org/newsletters/NL2009/Mar09.pdf.

27   Kim Heacox (ed.). *Visions of a Wild America*. Washington, D.C.: National Geographic Society, 1996, p. 20.

28   Mark Jacobson. "the hunt for red gold." *Onearth*: Fall 2004, p. 17.

29   Jack Kerouac. *On the Road*. New York: Penguin Books, 1955, p. 17.

Five to six million years of erosion have left the Grand Canyon a mile deep, 4 to 18 miles wide, and 217 miles long. Theodore Roosevelt called the Canyon "one of the great sights which every American if he can travel at all should see." John Burroughs wrote, "Words do not come readily to one's lips, or gestures to one's body, in the presence of such a scene. One of my companions said that the first thing that came into her mind was the old text, 'Be still, and know that I am God.'"[30]

By God, I thought after having experienced both Yosemite and the Grand Canyon, *there is no doubt that we truly do live on an exquisitely beautiful planet.* Reluctantly leaving them both behind, I reported for duty at Grand Forks Air Force Base with impressions of gaping redrock canyons and Giant Sequoias lodged firmly in my heart and mind. I didn't know when, but I'd be back.

Assigned to GFAFB's 446th Strategic Missile Squadron, part of the 321st Strategic Missile Wing, I completed my four-year service commitment as a missile launch officer and combat crew commander, not returning to the towering buttes, sweeping plateaus, and slender slot canyons of the Southwest or silent sentinels of the Sierras until after leaving the Air Force four years later.

VAGABOND

In the words of Edward Abbey, there comes a time in the life of us all when we must lay aside our books and put down our tools, leave our place of work, and walk forth on the road to meet the enemy face-to-face, once and for all and at last.[31] In late April 1995, I left my place of work, scuttling a moderately promising Air Force career, and traveled for the better part of two years, dropping everything in the name of adventure.

I'd been looking forward to this early, albeit temporary, "retirement" for nearly four years. Not long after reporting for active duty, I started contemplating completing four years of military service and then leaving the Air Force to travel the country and the world for a year or two. Just getting in and already planning on getting out, I was learning that life is more about experiences than careers, and living a life without risk is not one worth living.[32]

I saved and invested as much as possible while serving our country during those four years, then sold, gave away, or stored with friends and family all my minimal worldly possessions, and at the age of 27, was completely free of all social, and to a lesser extent, economic, responsibilities and obligations. I was

---

30   David Harmon (ed.). *Mirror of America*. Boulder, CO: Roberts Rinehart Publishers, 1989, p. 71.

31   Edward Abbey. *A Voice Crying in the Wilderness (Vox Clamantis in Deserto)*. New York: St. Martin's Press, 1989, p. 48.

32   See articles: David A. Lien. "Last Alert." *Association of Air Force Missileers (AAFM) Newsletter*: December 2005, p. 3 (http://www.afmissileers.org/newsletters/NL2005/Dec05.pdf).

ready to travel and explore, to experience life unimpeded by people or possessions, and it would be nearly four years before I'd return to the world of work.

I didn't have any plans or worries about post-Air Force employment at the time, much to the chagrin of my grandparents and some other family members, but what I had was plenty of time to do as I wished and to find out who I was while seeking out the heart and soul of this great nation and continent; the remaining wild places and wide-open spaces still mostly "untrammeled by man." I was living for myself and the moment, something some people talk about and many contemplate, but few have the determination or courage to try.

I traveled across North America from Alaska to Mexico, California to Florida, New Mexico to New Jersey, and learned in a way that only unfettered travel allows for while searching (literally and figuratively) for the meaning of life and experiencing true freedom. A freedom defined by expanding one's horizons beyond the life-stifling world of careers, car payments, mortgages, and 40-hour weeks. A life most Americans endure and embrace to the exclusion of almost all else. Wilderness Society founder Bob Marshall knew there is something glorious in traveling beyond the ends of the earth, in cutting loose from the bonds of civilization.

He may have been the last American able to dream of being Lewis and Clark and then find a blank space on the map (in Alaska) to explore.[33] As far as anyone knows, the first nonnative to explore central Alaska's Brooks Range was Marshall, who ventured into the north fork drainage of the Koyukuk River in 1929. It was a place, Marshall wrote, "two hundred miles beyond the edge of the Twentieth Century."[34]

Bob Marshall "cut loose" from civilization by exploring the wilds of Montana, Idaho, and the Arctic regions of Alaska during the 1920s and 1930s. I did it by traveling across the North American continent during the 1990s. A modern-day nomad, I lived and did as I wished with very few cares or concerns beyond the hour at hand. Bob Marshall and Henry David Thoreau would have approved of my vagabond wanderings.

Thoreau said, "If a man does not keep pace with his companions, perhaps it is because he hears a different drummer. Let him step to the music which he hears, however measured or far away." Thoreau's love of nature and solitude, and his simple, frugal lifestyle—designed to avoid a life of "quiet desperation"—struck a chord with me, and I sought to follow his advice for achieving one's dreams: "If one advances confidently in the direction of his dreams and endeavours to live the life he has imagined, he will meet with success unexpected in

---

33   Roderick Nash. *Wilderness And The American Mind*. Yale University Press: 1967, p. 340.

34   Rachel Carley. *Wilderness A to Z*. New York: Simon & Schuster, 2001, p. 52.

common hours."[35]

It was a glorious time and an incredible adventure, and I expect to live like that again someday, and hope some of you do too. Like Peter Matthiessen (author of *The Snow Leopard*) experienced while tracking those ghostlike cats in the mountains of Nepal, "With the past evaporated, the future pointless, and all expectation worn away, I began to experience that *now* that is spoken of by the great teachers."[36] If you ever consider doing something like Bob, Henry, Peter, and I did, or possibly a shorter version of it, and keep waiting until "someday," you never will. *Now* is the time. Life is precious, time is short. Life should be a trip, not a trap!

Northern Minnesota writer Jim Dale Huot-Vickery says, when, as adventurers, we pass through landscapes, we move through a cacophony of sensations. "There is sometimes risk and danger as flesh dances with ecosystems and climates," he wrote, "but the fortunate adventurer returns home a different person, subtly better, forever branded, having been worked upon. Impressed." Whether we are alert adventurers or people who awaken to and dwell in one landscape, "we enter a chrysalis and emerge, like Thoreau from Walden Pond, more perfect creatures."[37]

Naturalist Jennifer Ackerman adds, "Gaining deep familiarity with a landscape other than your native one is like learning to speak a foreign language... Slowly the strange becomes familiar; the familiar becomes precious." Because adventurers, travelers, explorers, hunters, trappers, hikers, climbers, and other outdoorsmen and women (Thoreau's "more perfect creatures") are routinely out in the still-wild places and wide-open spaces of the country and world, seeing and experiencing how incredible and necessary they are, and how rapidly they're disappearing, we have a responsibility to report to others the breadth and depth of what's being lost.

As David ("Thoreau with a Bow") Petersen says in his book *Elkheart*, "To criticize the bad is our duty to the good, eh, Edward?"[38] Ed would know. Writer and wildlands activist Edward Abbey (the "Thoreau of the American West") postulated that truth is the enemy of power, as power is the enemy of truth, and he explored the writer's "duty to speak the truth—especially unpopular truth. Especially truth that offends the powerful, the rich, the well-established, the traditional, the mythic, the sentimental. To attack, when the time makes it necessary, the sacred cows of his society."[39]

---

35   Robert Behnke. "The Wisdom of Theodore Gordon." *Trout*: Spring 2009, p. 58.

36   Peter Matthiessen. *The Snow Leopard*. New Delhi: Rupa & Co., 1978, p. 274.

37   John Henricksson (ed.). *North Writers II*. Minneapolis, MN: University of Minnesota Press, 1997, p. 59.

38   David Petersen. *Elkheart*. Boulder, CO: Johnson Books, 1998, p. 167.

39   Kim Heacox. *Visions of a Wild America*. Washington, D.C.: National Geographic Society, 1996, p. 148.

Abbey lived by Walt Whitman's credo: "Resist much, obey little." "It is my belief," Ed wrote in his essay *A Writer's Credo*, "that the writer... should be and must be a critic of the society in which he lives." In the following pages, I too will be a critic. I will tell the truth and attack the sacred cows of society: growth, capitalism, careers, civilization, government, trophy homes, reality TV, ATVs, RVs, SUVs, etc.

It's oftentimes a combination of these societal sacred cows that keep too many of us from appreciating and experiencing the real world, the wild and natural one we're so ruthlessly and efficiently liquidating in the name of growth, progress, and political payoffs. That said, my goal here is a simple one: to introduce you to some of the wild places I have experienced and explored over the years and to expose those who are destroying them, so that you might decide to help save the wildlands that remain.

Essayist Edward Hoagland says, "We are destroying God's Earth. It's just going before our eyes, and some of us who write about nature are so dumbfounded we are hardly able to articulate what we're seeing."[40] I'd like to think my upbringing, my years of higher education (three college degrees) and work (two careers: one military and one civilian), combined with frequent worldwide travel (28 countries so far) and climbing and other adventures have given me a broad perspective on life, society, civilization, and the wild and natural world.

I hope my life experiences have made me a more enlightened, "more perfect creature," and I think that humanity is at a tipping point. I believe we must come to the collective realization that the wild and natural world—creation itself—is the only thing we must save if we are to save ourselves. My opinions and their accompanying facts have some bite, you'll see, but there's no point in being timid. Anyone who portends to write anything worthwhile without ruffling some feathers probably doesn't have much to say.

"The only true dignity of man is his ability
to fight against insurmountable odds."
–Ignazio Silone

CROSS-COUNTRY CAREER SEARCH

During the fall of 1996, I reluctantly relinquished much of my newfound freedom—after traveling the country and (to a lesser extent) the world nearly non-stop since leaving the Air Force in April 1995—and started considering what I would eventually do for work, another career. My year-plus of vagabond traveling

---

40   The Wilderness Society. "Essayist Has Seen—and Not Seen—It All." *Wilderness*: Winter 2006–2007, p. 46.

"flew by like a drunk Friday night," and the party was coming to an end.

Career. The word itself is unappealing and conjures up images of toiling away at some miserable, meaningless job—which most of us do without a second thought—while our marginally satisfying, mostly unfulfilled lives slip slowly away. Alan Jackson got it right when he sang about "getting paid by the hour and older by the minute."[41] Before we know it, we're sitting in lawn chairs in our driveways wondering where all the years went and why we didn't do more with them.

Even though I would have preferred continuing my short career as a professional traveler, wanderer, and explorer (a modern-day nomad), I didn't feel like reverting to living under bridges and panhandling on street corners to maintain my career-free and carefree existence—even vagabonds need money to get by in our society. However, I also didn't want to endure an unsatisfying job just to pay the bills. Those vocations are a dime a dozen. I wanted to do something that meant more than just a paycheck. A tall order these days, but I was determined to try.

Having become a fairly serious investor and novice stock picker during the early 1990s—good years for convincing anyone they could profitably pick and flip stocks—I thought being a stockbroker might be a good career choice. What appealed to me most about the financial services industry was being in a position to help people achieve their financial goals. After researching the major retail brokerage firms and their training programs, and talking to a few practicing brokers, I decided Merrill Lynch had the best new broker training program.

During October 1996, Merrill's national recruiter invited me out to the company headquarters in Princeton, New Jersey, for their Financial Consultant Career Day, where they put prospective recruits through a company orientation and screening. Merrill was looking to hire former military officers, so their recruiter was enthusiastic about my visit. The perfect excuse for a road trip, traveling first across northern Wisconsin and Michigan's Upper Peninsula (the UP), then south over the Mackinac Straight Bridge into Michigan proper.

Before November 1957, explains *American Hunter* editor J. Scott Olmsted, getting to Michigan's Upper Peninsula meant a 45-minute ferry ride, not including time to queue and time to load. On busy weekends, like the start of hunting season, 24-hour waits in five-mile-long backups were not uncommon. Then, the Mackinac Bridge was completed, and crossing the four miles of the windswept straits that connect Lake Michigan and Lake Huron was reduced to a ten-minute car ride. Over the years, more than 100 million people—many of them hunters headed up to the U.P. every fall—have crossed the bridge.[42]

---

41   A song by Alan Jackson sung with Jimmy Buffett: "It's five o'clock somewhere."
42   J. Scott Olmsted. "10 Million Deer Hunters Roll Out for the Rut." *American Hunter*: November 2010, p. 31.

In Ann Arbor, I stopped to see an Air Force friend, Todd Laughman. Todd and I served together in the 446th Strategic Missile Squadron at Grand Forks Air Force Base. Like many missile launch officers and others forced to live in the semidesolation of eastern North Dakota, we (or maybe just me) developed a fairly cynical attitude about "the Dak," as Grand Forks was not so lovingly referred to by many missileers and others. Part of the problem was being stuck in the either bitter cold or the baking heat of the northeastern Great Plains with not much to do for engaging entertainment. "Disneyland for drunks" some called it.

> "We take no sides
> We take no prisoners
> We miss no Happy Hours."
> —M. John Fayhee[43]

In addition to a shared affinity for beer, Todd and I also independently developed a penchant for canoeing, camping, and hiking in the Boundary Waters Canoe Area Wilderness (BWCAW) of northern Minnesota, one of the Midwest's finest and wildest locales. Over the years, I've found that the bonds of camaraderie based on wilderness and similar outdoors experiences are generally the strongest and longest-lived. Even though I've only seen Todd a handful of times over recent years, we still keep in touch and he's still exploring and experiencing the wilds of the BWCAW and elsewhere when time permits.

After two nights of bar-hopping and beer drinking with Todd in Ann Arbor, I hit the road again and was particularly impressed by the rolling, hardwood-covered hills and valleys of Pennsylvania. The leaves weren't changing to their fall colors yet, but I'll bet it's a sight to see when they do. Nineteenth-century American naturalist John Burroughs once described October as "the time of the illuminated woods." That surely would be Pennsylvania in a few short weeks, "and if I were a bird I would fly about the earth seeking the successive autumns," George Eliot wrote.

Pennsylvania's hilly, thickly forested terrain is perfect country for white-tailed deer, and (coincidentally) deer hunting is what Pennsylvanians do, notes *American Hunter* contributor Kyle Wintersteen. The state's 950,000 hunters (only Texas has more) bag 250,000 to 500,000 deer annually. Its most heavily enrolled university, Penn State, has two fraternities devoted to hunting. The state animal? A whitetail deer. And Pennsylvania's first deer-hunting law was written 55 years before the Declaration of Independence.[44]

43   M. John Fayhee. *The Mountain Gazette #97.*
44   Kyle Wintersteen. "Classic Country: Pennsylvania." *American Hunter*: September 2009, p. 82.

Continuing on to New Jersey, where my first stop was Liberty State Park, I rode the last boat of the day out to Liberty Island. The Statue of Liberty was not nearly as tall as I'd imagined, always being portrayed as larger-than-life on television and in the movies. Nonetheless, it struck an impressive pose on its high pedestal in the harbor with Manhattan's skyline as a backdrop. The World Trade Center towers were still there then, unmistakable symbols of America's worldwide economic dominance in the post-Cold War era. However, they were part of a skyline that was destined to change dramatically on September 11, 2001.

Although their eventual destruction in the kamikaze-like attacks of 9/11 was horrific due to the sickening loss of life, the towers themselves (until that fateful day) were merely economic symbols to most Americans, and economics is not what made America great, it only made us collectively rich and militarily powerful. Some would say that is the definition of greatness. I do not. Freedom and democracy nurtured by over 200 years of wilderness experiences are what made and make America the greatest (still, in my opinion) country on earth.

And as I recall, the massive Twin Towers dominating Manhattan's skyline were completely overshadowed by that simple symbol of liberty standing alone in the harbor. Unlike with some things, when it comes to freedom and democracy, size doesn't matter. I took a picture or two of the Twin Towers, but ended up with a dozen or more of Lady Liberty. Solemn and humbled, I left her side and headed south for Princeton and Merrill Lynch.

> "The... job was exciting and challenging. I was very well paid. But one day I found myself on *The Donahue Show* hyping a course on 'How to Marry Money,' and I realized I had gotten off the path."
> —Jeffrey Hollender[45]

Over the next two days, six of us stockbroker wannabes attended Merrill Lynch's Career Day orientation, which included two interviews with the national recruiter, Chuck. During the first round of interviews Friday afternoon, Chuck asked me why I wanted to work for Merrill Lynch, why I wanted to be a broker. I replied truthfully, saying: "I want to make a difference in people's lives, I want to help them achieve their financial goals."

I was somewhat surprised when he looked me in the eye and said, "Wrong answer." The correct answer is, "You want to make money." Chuck explained (although I knew it already) that the brokerage industry is geared, quite simply and greedily, toward making money (like all capitalistic enterprises), and the kind

---

45 Jeffrey Hollender, president of Seventh Generation. *Onearth*: Winter 2004, p. 37.

of people it attracts are generally very competitive in that regard. As Edward Abbey said, "One thing more dangerous than getting between a grizzly sow and her cub is getting between a businessman and a dollar bill."[46]

Consequently, brokers spend the majority of their days prospecting for new clients, constantly looking to grow their portfolios, always searching for ways and angles to accumulate more clients and assets under management. Which is how they earn bigger commissions, buy larger houses, take more expensive vacations, and attract better looking spouses (the skin deep, superficial kind) who will demand that their supersized, overpriced homes be filled with more mostly useless and overpriced clutter. Again, from Ed Abbey: "The rich can buy everything but health, virtue, friendship, wit... love, pride, intelligence, grace, and, if you need it, happiness."[47]

On Saturday, we went through the rest of Merrill's orientation and I was impressed with the firm's training program and broker support, but also realized—being a person who is only motivated by money to a fairly limited extent—I probably wouldn't have the drive to be successful in the brokerage industry. Besides, I wasn't overly interested in being a glorified salesman, which is what a retail brokerage job boils down to.

At the conclusion of my second interview, Chuck said they had a place for me at Merrill, but asked me to think it over and give him a call later that week. He was straight with me, which I respected, and I was equally honest with him. Stopping off at a payphone on the way home Sunday afternoon, I called and thanked Chuck for his offer, but wasn't interested in being a broker anymore. There's a lot more to life than making money, and I'd have to find some less offensive, more rewarding way to do it.

I decided long ago to follow the advice and guidance of people like Edward Abbey and Edmund Burke when it came to what I'd be willing to do to earn a living. Some 250 years ago, Burke said: "The great Error of our Nature is not to know where to stop, not to be satisfied with any reasonable Acquirement; not to compound with our Condition; but to lose all we have gained by an insatiable Pursuit after more."[48] Heading northeast out of New Jersey, I drove through Pennsylvania into New York, and stopped at the breathtaking tourist trap that is Niagara Falls.

While serving as governor of New York (1898-1900), Theodore Roosevelt inspected Niagara Falls (explains Douglas Brinkley in *The Wilderness Warrior*)

---

46   Edward Abbey. *A Voice Crying in the Wilderness (Vox Clamantis in Deserto)*. New York: St. Martin's Press, 1989, p. 97.

47   Edward Abbey. *A Voice Crying in the Wilderness (Vox Clamantis in Deserto)*. New York: St. Martin's Press, 1989, p. 98.

48   David Jenkins. "Biting the Hand That Feeds Us." *C.E.P. Quarterly*: Spring 2007, p. 4.

to see if it could become a national park. The Transcendentalist philosopher Margaret Fuller once said the great falls were "the one object in the world that would not disappoint." Roosevelt disagreed. According to Brinkley, "In the spirit of Ripley's Believe It or Not, there were already five-legged calves and two-headed goats on display near the falls. Roosevelt dropped the issue—the concessionaires had already seized Niagara Falls, and there was no turning back."[49] After a brief stop at the falls I drove through Toronto, then along the north shore of Lake Huron, and camped at Six Mile Lake Provincial Park Sunday night.

On Monday, I returned to the U.S. through Sault Ste. Marie and Michigan's Upper Peninsula, a place that's home to a growing population of wolves, like my native Minnesota. In 1974, when *Canis lupus* was declared endangered in the Lower 48 states, the gray wolf population was confined to a corner of northern Minnesota in the Superior National Forest and Michigan's Isle Royale National Park in Lake Superior.[50] They've since returned to the UP, and "one cannot help fancying that he has gone to the ends of the earth, and beyond" when hearing their howls.[51]

Any place wild enough for wolves or grizzlies, but preferably both, is a some-place worth exploring. Ernest Hemingway thought so too. During the summer of 1919, Hemingway and two friends went on a fishing trip to the UP. They caught 200 trout in a week's time and Ernest, then 20, came away with the inspiration for one of his most famous stories, the "Big Two-Hearted River." Like Hemingway, I was captivated by the UP's forests, lakes, streams, wolves, and their collective call of the wild, reminding me that I was nearly home.

It's only in such places that true peace of mind, contentment, and "grace" can be found. In *A River Runs Through It*, Norman Maclean wrote, "All good things—trout as well as eternal salvation—come by grace and grace comes by art and art does not come easy." That night I camped in a graceful place, Tahquamenon Falls State Park (the second-largest state park in Michigan), a Northwoods haven encompassing 45,000 acres of the UP's wild public lands.

Tahquamenon Falls has more than 40 miles of hiking trails, 13 inland lakes, 24 miles of the Tahquamenon River, and 20,000 acres of designated natural areas. A place worthy of much more time than I was prepared to spend after having done so much traveling and exploring over the summer and fall. I left Tahquamenon Falls Tuesday morning and returned home to Minnesota and Grand Rapids that afternoon.

---

49 Douglas Brinkley. *The Wilderness Warrior: Theodore Roosevelt And The Crusade For America*. New York: HarperCollins Publishers, 2009, p. 353.
50 Douglas H. Chadwick. "Wolf Wars." *National Geographic*: March 2010, p. 38.
51 Jeff Rennicke. "Keweenaw National Historic Park." *National Parks*: Winter 2010, p. 28.

I had traveled 3,450 miles over 10 days at a cost of $650 to find out I didn't want to be a stockbroker and now needed a new career idea. Not being in any real rush to return to the world of career-driven living, I spent the rest of the fall hiking, camping, grouse and deer hunting, and generally enjoying the best time of the year in northern Minnesota. Then, in November, I registered for four winter quarter accounting and business classes at Itasca Community College (ICC) in Grand Rapids. After completing winter quarter, I moved (back) to Duluth, Minnesota.

If I ever live in northern Minnesota again and end up in a sizable city, Duluth will likely be it. Sometimes called the "San Francisco of the North," Duluth is situated at the crossroads of the best of northern Minnesota. Sitting high atop a hill overlooking Lake Superior, Duluth and the University of Minnesota-Duluth (UMD) is a good combination for anyone interested in getting a fine education and having excellent access to the outdoors.

In 2004, *Paddling* magazine named UMD one of the top 10 paddling colleges in the nation, and *Outside* magazine ranked it as one of the 40 best college towns in North America.[52] In addition, Duluth came in 92nd among the best towns in America for hunters and anglers, according to the June/July 2009 issue of *Outdoor Life*. Lewiston, Idaho, ranked first and Marquette, Michigan, second.[53]

Among other towns in the top 20 were Bismarck, N.D.; Pierre, S.D.; Rochester, Minnesota; and Roseau, Minn. Towns were ranked based on more than 20 criteria, with outdoor-related factors given slightly heavier emphasis in computing the results than quality-of-life factors, including gun-friendliness of each town's state, huntable and fishable species nearby, proximity to public hunting land and fishable waters, and the potential for taking trophy-caliber game or fish.[54]

Not to be outdone, during 2009, *Outside* magazine ranked Colorado Springs (my current home) as the number-one city to live in. As explained by Colorado Springs *Gazette* contributor Scott Rappold, there's a reason 1.3 million people have rushed to Colorado's Front Range in the past two decades: with 249 annual days of sun, an ascendant, heavily tech-based economy, and quick access to nearly four million acres of Rocky Mountain wilderness and a dozen world-class ski resorts, "it's simply a pretty awesome place to live."[55]

People also come to Colorado Springs, Scott adds, for 14,110-foot Pikes Peak (towering above town); the Arkansas River's Class IV rapids (two hours west); world-class athletic facilities (Carmichael Training Systems is based here);

---

52   University of Minnesota-Duluth (UMD). *UMD Momenta*: Winter 2004, p. 1.

53   Sam Cook. "We're No. 92!" *Duluth News Tribune*: 6/9/09.

54   Sam Cook. "We're No. 92!" *Duluth News Tribune*: 6/9/09.

55   Scott Rappold. "They like us! They really like us!" *The* [Colorado Springs] *Gazette*: 7/28/09.

and 260 miles of multisport trails available within a 10-mile radius. "Sure," he notes, "there are a few other towns with this many outdoor options, but they generally cost twice as much."[56] Bottom line: both Duluth and Colorado Springs are pretty cool places to set up camp.

UMD ᴀɴᴅ NCUA

After two years of recurring travel (i.e., real-world education) in the U.S. and abroad, during the spring of 1997 I returned to UMD to finish my formal collegiate education. In 1990, I'd earned an undergraduate degree in political science there. This time, it was an accounting degree, which took about a year-and-a-half to complete. I guessed correctly that a bachelor of accounting combined with a master of science in administration, coupled with my military background, would provide the necessary credentials for another government job of some sort.

I also knew that by continuing to work for the federal government, my time in the Air Force would count toward seniority, pay, retirement, and most importantly, annual leave. While perusing government job openings on the Internet (at *www.USAJOBS.gov*), I found one in Colorado Springs working as a credit union examiner for the National Credit Union Administration (NCUA). The NCUA is an independent financial regulatory agency of the federal government responsible for chartering, supervising, examining, and insuring all federal and most state-chartered credit unions.

I didn't know much about credit unions at the time, but was pleased to find out they're not your typical greed-driven financial institutions. While almost every business relationship is capitalistic in nature—where businesses are compelled to extract as much as possible from their customers to placate shareholders and overcompensated CEOs and boards—credit unions function in a symbiotic environment, explains *Credit Union* magazine contributor Lisa McCue. "As not-for-profits, credit unions have no deep pockets demanding to be filled," she says, "and no obligation to show off record earnings or to please stockholders through dividends. Credit unions look out for the little guy."[57]

Bill Raker, president/CEO of U.S. Federal Credit Union in Burnsville, Minnesota, says credit unions are "financial institutions with a heart," while other financial institutions just have a bottom line. "We're simply doing what credit unions have always done for their members," he explains. "Reaching out to individuals and families to help them achieve their financial dreams."[58] A perfect

56 Scott Rappold. "They like us! They really like us!" *The* [Colorado Springs] *Gazette:* 7/28/09.

57 Lisa J. McCue. "It's Time to Be Counted." *Credit Union* magazine: March 2008, p. 44.

58 Dick Radtke and Lucy Harr. "Icing on the Cake." *Credit Union* magazine: May 2008, p. 28.

match for me and my modestly lofty ideal of doing something more with a career than just earning a fat paycheck, which seems to drive the career decisions of far too many people in our increasingly materialistic, consumption-crazed society.

A federal law passed in 1934–a year in which U.S. unemployment stood at 22 percent and dust storms in the Central Plains created the "Dust Bowl"–is behind much of the business-as-usual of today's credit unions. The 73rd Congress passed the Federal Credit Union Act on June 16, 1934, and President Franklin D. Roosevelt signed it into law 10 days later. The act emphasizes the nation's credit needs:[59]

> "An Act to establish a Federal Credit Union System, to establish a further market for securities of the United States, and to make more available to people of small means credit for provident purposes through a national system of cooperative credit, thereby helping to stabilize the credit structure of the United States."[60]

Edward Filene founded the U.S. credit union movement, and the Federal Credit Union Act was a critical step toward fulfilling Filene's vision of a financial system allowing average people to control their financial destiny.[61] Although the credit union examiner job sounded like quite a leap from anything I'd done before, it turned out to be a perfect match, and the opening was in Colorado Springs.

Several of my career Air Force friends were stationed in Colorado Springs and Denver, and I wanted to explore the mountains and canyon country of the Southwest some more, so I applied for the job, not caring too much about what being a credit union examiner actually involved. At the end of the day (in the words of John Muir), "anyplace that is wild" is where I wanted to be, and in Colorado there are 3.7 million wilderness acres and over 4 million acres of Inventoried Roadless Areas to experience and explore.

Wallace Stegner would have liked my plan. "It should not be denied," he wrote, "that being footloose has always exhilarated us. It is associated in our minds with escape from history and oppression and law and irksome obligations, with absolute freedom, and the road has always led west." *Anchorage Daily News* columnist Craig Medred says that almost from the beginning our modern

---

59 Credit Union National Association (CUNA). "Federal CU Act Passes: 1934." *Credit Union* magazine: December 2008, p. 64.

60 Credit Union National Association (CUNA). "Federal CU Act Passes: 1934." *Credit Union* magazine: December 2008, p. 64.

61 Credit Union National Association (CUNA). "Federal CU Act Passes: 1934." *Credit Union* magazine: December 2008, p. 64.

ancestors pushed relentlessly west on foot, by boat, by cart, or with wagons. These were the first great road trips in the modern sense, and North Americans have been road-tripping ever since.[62]

But before beginning my westward journey, I still had to land that examiner job by jumping through some hoops established by the Office of Personnel Management (OPM). The OPM is a federal government agency that serves as a gatekeeper for most civilian government jobs. After being deemed qualified by the OPM and given a score based on my education, previous military experience, and other marketable skills, my application package was forwarded to the NCUA and supervisory examiner Lisa Dolin.

Lisa flew me from Duluth to Denver for an interview at the Denver airport. We talked for about forty-five minutes, then I returned to Duluth that afternoon. A few days later, I received a call from the NCUA's Region 5 office in Austin, Texas, with a job offer. I also had offers from Honeywell in Minneapolis and a possible job with the state of Minnesota's financial regulatory division, but I declined that interview due to having already accepted the credit union examiner position in Colorado Springs.

As an examiner, I would (for starters) be contributing to maintaining the solvency of the nation's financial system by helping to ensure that credit unions are operated in a safe and sound manner. The NCUA, with the backing of the full faith and credit of the U.S. government, operates and manages the National Credit Union Share Insurance Fund (NCUSIF), insuring the deposits of 90 million account holders in credit unions nationwide. And in the 75-year history of credit unions, they've never cost taxpayers a dime.

Notwithstanding my career-based selflessness, this job could not have come at a better time financially. I was basically broke, as you might expect, after what had turned into over three years of travel, school, and joblessness—some of the best years of my life. In addition, my Chevy Blazer's engine blew that fall and my credit card debt grew to the tune of around $4,000 as a result. I'd had a good run, but it was time to return to the world of work. It was now the late fall of 1998.

For some historical perspective, you might remember that 1998 was the year John Glenn returned to space, Jesse Ventura became the governor of Minnesota, Google, Inc. was founded in California, the Denver Broncos won their first-ever Super Bowl, and I moved from Duluth, Minnesota, to Colorado Springs, Colorado.[63] This move was the culmination of a three-and-a-half year transition period in my life: two years as a vagabond and a year-and-a-half as a

---

62   Craig Medred. "Road trip! Answering a genetic call." *Anchorage Daily News*: 9/17/06.

63   Jeff Parsons. "Ten years of Protecting Colorado Landscapes." *Colorado Wild*: Winter 2008-2009, p. 1.

full-time student vagabond.

I arrived in Colorado Springs during November 1998, started working for the NCUA on January 3, 1999, and climbed my first fourteener (Longs Peak in Rocky Mountain National Park) on August 14, 1999. Three summers later, I would climb my 44th fourteener, Mount Wilson (on July 5, 2002), and four years and three days later (on August 17, 2003), my 55th fourteener, Mount Eolus.

> "I reckon I got to light out for the territory ahead
> of the rest, because Aunt Polly is tryin' to civilize
> me. I've tried it, and it don't work."
> —Huck Finn[64]

COLORADO AND THE ROCKY MOUNTAINS

The Rocky Mountains, explains Rachel Carley in *Wilderness A to Z*, extend from northwestern Alaska's Brooks Range, through Canada's Yukon Territory, down into Washington, Idaho, Montana, Utah, Wyoming, and Colorado, finally terminating in northern New Mexico. Along the way, the Rockies measure anywhere from 70 to 400 miles across as they rise from the Great Plains on the east and fall to the Basin and Range region to the west.[65]

Among the many mountain chains within the greater Rocky Mountain system are the Teton, Uinta, and Wasatch Mountains; the Absaroka, Bighorn, Bitterroot, Gallatin, Laramie, Salmon River, and Sawtooth Mountains; the Medicine Bow, Park, San Juan, Sangre de Cristo, Sawatch, and Wind River Ranges; and many smaller groups.[66]

Carley says our first real knowledge of the Rockies came during the era of the mountain men, which began with the pioneering fur expeditions of William Ashley (1778-1838) from St. Louis into the Yellowstone region in the 1820s and continued with successive fur-trading ventures until the beaver trade collapsed in the 1840s. During this period, an extended band of scouts worked deep into the Rockies, where such legendary figures as John Colter (1775-1813), Jedediah Smith (1799-1831), Jim Bridger (1804-81), and Kit Carson (1809-68) became familiar with existing Indian routes and began to break new trails to California and the Pacific Northwest.[67]

As of 1849, only five states had been laid out west of the Mississippi—Texas,

---

64   Mark Twain's Huck Finn, as quoted by Tom Anderson. "Northern exposure: Moving to the Yukon." [Minneapolis-St. Paul] *Star Tribune*: 12/27/08.

65   Rachel Carley. *Wilderness A to Z*. New York: Simon & Schuster, 2001, p. 263.

66   Rachel Carley. *Wilderness A to Z*. New York: Simon & Schuster, 2001, p. 263.

67   Rachel Carley. *Wilderness A to Z*. New York: Simon & Schuster, 2001, p. 264.

Louisiana, Arkansas, Missouri, and Iowa—and not one mile of railroad track. To determine the best route to the West Coast and thereby open up the west to commerce and settlement, the 1852 Pacific Railway Survey Act launched six surveys into the western territories in the mid-1850s.[68] The last four major western surveys began after the Civil War and were completed by 1879.[69]

As immigrants and Americans began the long journey westward in the 1870s, adds *Inside/Outside Southwest* contributor Andrew Gulliford (in "The Hayden Atlas turns 130 years old"), Congress funded four surveys of the west: by F.V. Hayden, John Wesley Powell, George M. Wheeler, and Clarence King. In 1880, the surveys were combined to become the U.S. Geological Survey.[70]

Historian Gil Mull, a private map collector and retired geologist from Santa Fe, New Mexico, knows a lot about the history of mapping the American West, and says, "Of the four surveys, the Hayden Survey was the largest and of longest duration. In addition to written reports, the Hayden Survey published separate topographic, drainage, and geologic maps for many areas in Utah, Wyoming, Montana, and Idaho, but only Colorado was mapped in its entirety."[71]

The Hayden and the Wheeler Surveys essentially shadowed each other across Colorado, sometimes devising names and calculating elevations that contradicted one another, explains Colorado Mountain Club archivist Woody Smith. Since members of the Hayden Survey had better political connections, their tabulations were favored and their place-names were almost always the ones that became official. How else could Hayden Survey topographer A. D. Wilson have *two* fourteeners (Wilson Peak and Mount Wilson) named after himself?[72]

After surveying the Yellowstone region and promoting it before Congress in 1872 (the year Yellowstone was designated a national park), Gulliford notes, Hayden set himself the nearly impossible task of surveying and naming the Colorado Rockies during summer expeditions in 1873, 1874, 1875, and 1876. Colorado was not yet a state. To obtain precise measurements, survey teams working under Dr. Hayden lugged 50-pound theodolites (combined transits and levels) to the summits of Colorado's principal peaks.[73]

Hayden's topographers included James Gardner, leading mapmaker and head topographer; A. D. Wilson, head of the Southwest Colorado survey team; Gustavus R. Bechler, head of the Yampa Division; George B. Chittendon, official topographer of the White River Division, and W. H. Holmes, topographer of the

68   Rachel Carley. *Wilderness A to Z.* New York: Simon & Schuster, 2001, p. 282.

69   Rachel Carley. *Wilderness A to Z.* New York: Simon & Schuster, 2001, p. 283.

70   Andrew Gulliford. "The Hayden Atlas turns 130 years old." *Inside/Outside Southwest:* April/May 2007.

71   Andrew Gulliford. "The Hayden Atlas turns 130 years old." *Inside/Outside Southwest:* April/May 2007.

72   Woody Smith, CMC archivist. "Place-name polka: Humboldt Peak." *Trail & Timberline:* Winter 2007- 2008, p. 41.

73   Andrew Gulliford. "The Hayden Atlas turns 130 years old." *Inside/Outside Southwest:* April/May 2007.

Dolores Canyon and San Miguel and Bear River Mountains. All of Hayden's men found mapping Colorado a magnificent professional challenge, and within three years, they had covered 70,000 square miles of rugged mountain country.[74]

Hayden encouraged his surveyors to strive for accuracy. To gauge the proper elevation of a 14,000-foot peak, a survey team had to spend the night above timberline and break camp before dawn to reach the summit in time to meet the sun's earliest rays. Three times (on August 20, 1873) surveyor James Gardner's assistant poked him in the back with a transit tripod to keep him from falling backwards off Snowmass Mountain. One party, attempting to scale a 13,000-foot peak, chose the potentially perilous route of following fresh grizzly bear tracks to reach the summit.[75]

As huge cumulonimbus thunderclouds moved across mountaintops during summer afternoons, lightning storms were serious threats to the surveyors with their metal tripods perched always at the highest points. A. D. Wilson wrote that after finally reaching the summit of a peak, he and his partner began to feel "a peculiar tickling sensation" along the roots of their hair. It was accompanied by "a peculiar sound almost exactly like that produced by frying bacon." As the thunderheads swept closer, the surveyors' transit began to click like a telegraph and their hair stood on end.[76]

"By this time," Wilson's companion remarked in an obvious pun, "we were electrified, and our notes were taken and recorded with lightning speed... When we raised our hats our hair stood on end, the sharp points of the hundred stones about us emitted continuous sounds, while the instrument out sang every-thing else." They dashed off the peak and were 30 feet below the summit when lightning struck.[77] As these early surveyors discovered, Colorado encompasses a diverse and potentially dangerous region of mountains, plateaus, canyons, and plains.

Generally, the eastern half of the state has flat, high plains and rolling prairies gradually rising westward to the Front Range foothills and to the higher ranges of the Rocky Mountains. The Continental Divide runs from north to south through west central Colorado and bisects the state into the eastern and western slopes. The western half of the state consists of alpine terrain interspersed with wide valleys, rugged canyons, high plateaus, and deep basins.[78]

Colorado's altitude is one of its distinctive geographical features, making it on average the nation's highest state: the average elevation is 6,800 feet. The

74   Andrew Gulliford. "The Hayden Atlas turns 130 years old." *Inside/Outside Southwest*: April/May 2007.
75   Andrew Gulliford. "The Hayden Atlas turns 130 years old." *Inside/Outside Southwest*: April/May 2007.
76   Andrew Gulliford. "The Hayden Atlas turns 130 years old." *Inside/Outside Southwest*: April/May 2007.
77   Andrew Gulliford. "The Hayden Atlas turns 130 years old." *Inside/Outside Southwest*: April/May 2007.
78   Colorado State Archives.

state's lowest point is at 3,315 feet on the Arikaree River where it flows into the northwestern corner of Kansas. Colorado's highest peak is Mount Elbert, at 14,431 feet, or 2.72 miles above sea level. Elbert is also the 14th highest peak in the United States, including Alaska, and the second highest in the Lower 48 states.[79]

All of Colorado's 14,000-foot peaks are found in the heart of America's wildest and most rugged mountains, the Rockies. The first Europeans to approach these high peaks from the east were French traders and trappers. Their preferred name, *Montagnes de Pierres Brilliantes* ("Shining Mountains"), gave way to Lewis and Clark's "Rocky Mountains" following the 1803 Louisiana Purchase.

In "My Year With Lewis & Clark," *National Geographic* contributor Anthony Brandt says during the early 1800s, the best geographers of the day held that the rumored Rocky Mountains were no higher than the Blue Ridge Mountains of Virginia and constituted not ranges and ranges of peaks, but a single ridgeline. You can imagine the Lewis & Clark expedition members' astonishment when they saw what the Rockies really were: unimaginable miles of interlaced ranges and precipitous peaks stretching for as far as the eye can see, and covered with snow. They had likely never seen mountains snow-covered in early summer.[80]

In "The Backbone of the World," *Greater Yellowstone Report* contributor Douglas Chadwick says the chain of peaks (i.e., the Rocky Mountains) stretching from New Mexico to Alaska's northern rim is the spine to which the continent's broad muscles are attached. Its grandeur divides the waters and feeds them freshly ground sediments. They shape the weather east and west, and within their contours is stored the biggest, shaggiest, wildest collection of free-roaming creatures left in North America.[81]

Coloradan Helen Hunt Jackson traveled extensively throughout the Rockies during the late 1800s. "To the outer edge of the concentric, curving ranges of this Rocky Mountain chain," she wrote, "one might journey, in and out and up and over, and in and out and in and out again, I am persuaded, all summer long, for summers and summers, and find no monotony, no repetition."[82] To quote poet Walt Whitman, "One needs new words when writing about these mountains and all the American West—the terms *far*, *large*, *vast*, etc., are insufficient."

The Rockies' vast and unsullied landscapes provided inspiration for some of the earliest wilderness preservation efforts. Yellowstone, the world's first national park, was established in 1872. Canada's Banff National Park came along 13 years later. The National Forest System's predecessor began in 1891 with the

79   Colorado State Archives.
80   Anthony Brandt. "My Year With Lewis & Clark." *National Geographic Adventure*: May 2003, p. 78.
81   Douglas Chadwick. "The Backbone of the World." *Greater Yellowstone Coalition 2002 Annual Report*, p. 18.
82   Frederick R. Rinehart (ed.). *Chronicles of Colorado*. Niwot, CO: Roberts Rinehart Publishers, 1993, p. 79.

creation of the Yellowstone National Park Timberland Reserve. The first success-ful effort to save an endangered species—the bison—was started in Yellowstone during the late 1890s.[83]

The Wilderness Society was founded in 1935 by eight distinguished con-servationists concerned about the rapid loss of roadless public lands and other wildlands in the Rockies and across the nation: Bob Marshall, Aldo Leopold, Robert Sterling Yard, Benton MacKaye, Ernest Oberholtzer, Bernard Frank, Harvey Broome, and Harold Clinton Anderson. Soon, Olaus Murie joined the leadership ranks; later, Howard Zahniser took over as executive director.[84]

All shared a vision of systematic protection of our nation's wild places and strove diligently throughout their lives to bring this vision to reality in the Rockies and elsewhere. They followed the sage advice of people like Ansel Adams, who said, "Let us leave a splendid legacy for our children... let us turn to them and say, this you inherit: guard it well, for it is far more precious than money... and once destroyed, nature's beauty cannot be repurchased at any price."

I first set foot in the Rocky Mountains of Colorado in the spring of 1985 during a high school ski trip to Steamboat Springs. With two dozen kids crammed into vans like cordwood, we drove straight through from Grand Rapids, Minnesota, stopping only briefly at places like Wall Drug, South Dakota, and other less no-table pitstops to air out and resupply. For some of my friends, this would be their first and last trip to the Rocky Mountains, but for me, it was just the beginning.

It was a miserable, cramped drive across the Great Plains, but I remember being amazed at first sight of that distant chain of mountains straddling the ho-rizon. With faces plastered against windows—necks, heads, and eyes crooked skyward—crossing through narrow canyons and over high passes, we entered the Front Range of the Rocky Mountains. A place surely not far from heaven for a bunch of high school ski team kids raised on the short, icy slopes of northern Minnesota.

Edward Abbey felt similarly when he first saw the Rockies: "Near Greybull I saw for the first time something I had dreamed of seeing for ten years. There on the western horizon, under a hot, clear sky, sixty miles away, crowned with snow (in July), was a magical vision, a legend come true: the Front Range of the Rocky Mountains. An impossible beauty, like a boy's first sight of an undressed girl, the image of those mountains struck a fundamental chord in my imagination that has sounded ever since."[85]

Although awestruck at first sight, I was downright exhilarated and bursting with anticipation preparing to ski down one of Steamboat's seemingly massive

83   George Wuerthner. "Keeping The Grizzly in Grizzly Creek." *The Wilderness Society*: 2001-2002, p. 14.
84   The Wilderness Society (TWS). *The Wilderness Act Handbook*. Washington, D.C.: TWS, 2004, p. 2.
85   Edward Abbey. *The Journey Home*. New York: Penguin Books, 1991, p. 2.

mountains for the first time. It was a feeling, a chord that struck deep into my soul, down to the essence of my very being. While it's rarely been repeated later in life, I still get a sense of it every time I set foot on any sky-scraping peak in the mountains of Colorado or any other wild place.

Anyone who has ever looked west across the Great Plains to a distant horizon broken by the purple flanks and white peaks of the Rockies likely remembers their first time. That image leaves an indelible imprint in your mind, and the same can surely be said for those encountering Alaska's Mount McKinley, Japan's Mount Fuji, Mount Kilimanjaro in Africa, and Mount Everest in Nepal for the first time. Powerful images nearly impossible to forget. Colorado, as a whole, has a similar affect on many people.

Renowned Rocky Mountain photographer John Fielder may be biased, but he believes no U.S. mountains can match the Colorado Rockies. The Sierra Nevada in California is certainly spectacular, he says, but too young to have Colorado's vast tundra. The Cascades of Washington and Oregon contain great forests and alpine scenery, but lack the clear blue skies and yellow autumns of Colorado. The northern Rockies of Wyoming and Montana boast serrated ridges and incredible networks of lakes and tarns, but like the Sierras, they don't have enough soil nor moisture to support the floral splendor of Colorado.[86]

It's not difficult to see why one can fall in love with this state, but not just because of its renowned fourteeners and countless other mountains. *Backpacker* magazine says Colorado is the best state for hikers, backpackers, climbers, hunters, campers, and other outdoorsmen and women because of its super-abundance of outdoor amenities, including:[87]

- 43 wilderness areas;
- 42 state parks;
- 54 14,000-foot peaks;
- 584 13,000-foot peaks;
- 3.7 million wilderness acres;
- 4.4 million roadless acres;
- 24.5 million acres of public land;
- 25 national parks/monuments/forests;
- 300-plus state wildlife areas;
- 8 national wildlife refuges; and
- 1,300 miles of long trail (i.e., the Colorado and Continental Divide trails)

---

86   John Fielder and Mark Pearson. *Colorado's Wilderness Areas*. Englewood, CO: Westcliffe Publishers, 1994, p. 5.

87   Backpacker magazine. "Hiker Nation: Rocky Mountain High." *Backpacker*: January 2009, p. 96.

The Continental Divide Trail (CDT) snakes along the Rockies from Canada to Mexico through the western states of Montana, Idaho, Wyoming, Colorado, and New Mexico. In Colorado alone, the CDT traverses 800 spectacular miles: across the alpine tundra of the South San Juan, Weminuche, and La Garita wilderness areas, past the remains of western settlements that supported 1800s gold and silver mining, through the Collegiate Peaks, Rocky Mountain National Park, and secluded Never Summer and Mount Zirkel wilderness areas.[88]

Elevations on the CDT are consistently higher than any other National Scenic Trail, with the low point being at 4,196 feet along Waterton Lake in Montana's Glacier National Park, and the highest point at 14,270 feet on Grays Peak, Colorado.[89] As S. F. Emmons (of the U.S. Geological Survey) wrote in 1883, "On all the broad extent of these United States, certainly no region can be found which presents more facts of interest, more opportunities for investigation, and greater possibilities of new discoveries, than the state of Colorado."

> "There's something ever egotistical in mountaintops
> and towers, and all things grand and lofty."
> —Herman Melville, *Moby Dick*

FOURTEENERS

Backpacker magazine contributor John Harlin explains (in "Rocky Mountain National Park") that for millennia in a land now known as Colorado, the wind blew and snow drifted. The crystals piled deeper and deeper until they compressed into rivers of ice that carved vast gashes into billion-year-old granite and gneiss. When the earth warmed and the ice melted, jagged spires and towering walls stood proud where they'd resisted the relentless ice flows.[90]

A myriad of high peaks and untold miles of mountainous terrain remained behind, making Colorado what it is today: a hiker's, backpacker's, climber's, hunter's, and angler's paradise. "Colorado is the undeniable apex of America, containing nearly 80 percent of the 68 fourteeners [be aware that counts vary] in the contiguous United States," says *Denver Post* outdoors columnist Scott Willoughby.[91] Rocky Mountain National Park, home to fourteener Longs Peak, is the nation's tenth national park and the epitome of this Coloradan landscape.

Rocky Mountain National Park was established in 1915 thanks, in part, to 25 people, who, in 1912, founded a small, pioneering club for exploring Colorado's

---

88   Lori Spaulding. "Trail challenges thru-hikers along the Continental Divide." *PikesPique*: September 2008, p. 1.
89   Glenn Scherer. "Walking Down a Dream." *American Hiker*: Winter 2008, p. 10.
90   John Harlin. "Rocky Mountain National Park." *Backpacker*: February 2003, p. 63.
91   Scott Willoughby. "Colorado's 14ers perfect for summit meetings." *The Denver Post*: 7/6/10.

mountains–it became known as the Colorado Mountain Club (CMC).[92] The club's fascination with 14,000-foot peaks is lost on much of the rest of the world, where the metric system reigns supreme. An equally magical number in the European Alps is 4,000 meters, which translates to 13,123 ft.[93]

Most of Colorado's fourteeners ("14ers")-officially there are 54, but the number ranges from 53 to 59, depending on who's doing the counting-can be climbed in a single day on their standard routes. Hence, they're becoming ever more popular as the population of Colorado and the Rocky Mountain West soars. While the U.S. Geological Survey (USGS) lists 53 distinct summits over 14,000 feet, some purists who count even subsidiary summits above that elevation come up with 59 peaks.[94]

The disparity arises due to differing fourteener definitions. At the most basic level, a fourteener is a natural geographical point with an elevation between 14,000 and 14,999 feet that descends in every direction. However, the Colorado mountaineering community uses an elevation gain of 300 feet from the saddle with the parent peak (the 300-foot rule) as a key indicator of each peak's significance. On top of that, difficulty, distance from the parent peak, and climbing heritage are taken into consideration.[95]

For example, although Conundrum Peak's elevation of 14,022 feet places it higher than eight other ranked Colorado fourteeners, it doesn't make the cut on most bonafide 14er lists because of its close proximity to Castle Peak and a less than 300 foot vertical drop to the connecting saddle.[96] The CMC's official list of 54 fourteeners includes 52 "hard-ranked" summits (i.e., peaks that rise at least 300 feet above *all* saddles separating them from other summits). Two additional peaks do not meet that criterion, but have historically been included on the list: North Maroon Peak and El Diente.[97]

The CMC list, however, does not include Challenger Point, which qualifies according to the 300-foot rule. Many climbers include it, along with Cameron, Conundrum, and North Eolus.[98] Differing fourteener definitions aside, 54 seems to be the generally accepted number of 14ers these days. I opted for 55, the number in the 1999 edition of Gerry Roach's *Colorado's Fourteeners: From Hikes to Climbs*–the climbing bible for many 14er aficionados.

With a climbing career spanning 50 years, world-class mountaineer Gerry Roach could be considered the ultimate peak bagger, says CMC member Lori

92   Gerald Caplan. "The Big Bang." *Trail & Timberline*: Fall 2009.
93   http://www.fourteenerworld.com/.
94   Lawrence Gonzales. "One Way Out." *National Geographic Adventure*: August 2003, p. 41.
95   http://www.thecavedog.com/14ers-Web_Pages/14ers-General_Statistics/14ers-General_Statistics.html.
96   http://www.fourteenerworld.com/.
97   Linda K. Crockett. "The 14er Files." *Trail & Timberline*: Winter 2007-2008, p. 15.
98   Linda K. Crockett. "The 14er Files." *Trail & Timberline*: Winter 2007-2008, p. 15.

Spaulding. He summited Mount Everest in 1983 and two years later became the second person to climb the highest peak on each of the seven continents, following in the footsteps of Dick Bass. He has traveled the world over, participating in numerous expeditions in Alaska, the Andes, and the Himalayas.[99]

Roach was also the first person to climb the 13 highest peaks in North America. In Colorado, he has climbed more than 1,500 peaks, including all of the 14ers. He's also the author of two autobiographies: *Transcendent Summits* (which covers how and why he started climbing) and *Ride the Breath* (detailing his classic climbs around the world).[100] The Colorado Springs *Gazette's* fourteeners guru, Dave Philipps, explains in "Fourteeners 101" that Colorado mountain guides are referred to by the author's name.

If someone mentions the Roach map to Pikes Peak, Dave notes, it's not a map showing where the bugs are, it's Gerry Roach's guide to the mountain—*Colorado's Fourteeners*. The most popular and up-to-date (according to Dave) of Colorado's fourteener guides, Roach's pairs detailed maps with consistent, accurate directions, and he throws in some helpful weather tips and hidden philosophical nuggets, known as "Roachisms," to break up the cut-and-dry text.[101]

Philipps also explains that most of Colorado's fourteeners "don't require tail-tucking technical moves. The summit routes are long, hard, and rocky—and the altitude can feel like a jackhammer on your head—but with the right preparation, most hikers can complete them."[102] In "Fourteeners challenge spirit, legs," *Rocky Mountain News* contributor Robert Weller says the growing popularity of fourteener climbing means they're beginning to show some wear and tear.[103]

Some of the peaks have become so popular that a group called the Colorado Fourteeners Initiative ("CFI": www.14ers.org) was formed in 1994 to enlist the help of volunteers who build trails and complete other projects to mitigate the damaging impacts from hordes of hikers and climbers on these high mountain environs.[104] I've trod upon plenty of CFI's work over the years, and it's definitely making a difference, protecting the 14ers from all of us. More on the CFI later.

Colorado's fourteeners are found in six mountain ranges: the Front Range includes Longs Peak, Mount Evans, Pikes Peak, and three others. The Mosquito Range, south of Interstate 70 near Breckenridge and Leadville, has five mostly gentle fourteeners, including Quandary Peak and Mounts Democrat and Lincoln. The Sawatch Range has 15 fourteeners, same as the entire state of California,

---

99   Lori Spaulding. "Roach finds adventure on peaks near and far." *PikesPique*: March 2010, p. 1.
100   Lori Spaulding. "Roach finds adventure on peaks near and far." *PikesPique*: March 2010, p. 1.
101   Dave Philipps. "Fourteeners 101." *The* [Colorado Springs] *Gazette*: 3/28/07.
102   Dave Philipps. "Fourteeners 101." *The* [Colorado Springs] *Gazette*: 3/28/07.
103   Robert Weller. "Fourteeners challenge spirit, legs." *Rocky Mountain News*: 7/14/01.
104   Robert Weller. "Fourteeners challenge spirit, legs." *Rocky Mountain News*: 7/14/01.

including Mount of the Holy Cross and the five Collegiate Peaks, which bear the names of well-known eastern universities.

A handful of Sawatch Range peaks have names honoring Ute Indians, who occupied this region for roughly six centuries before Euro-Americans arrived.[105] Ten fourteeners are in the Sangre de Cristo (or "Blood of Christ") Range in southern Colorado, and the six fourteeners in the Elk Range near Aspen are among the most rugged. Thirteen fourteeners are found in the San Juans, a remote 4,000-square-mile region in southwestern Colorado.[106]

Although the Sawatch Range contains the most fourteeners in Colorado, the San Juans have the largest number of centennial thirteeners (those ranging from 13,800 to 13,999 ft.).[107] Scott Willoughby says, "It may be the steep and dramatic San Juans bulging into the southwestern portion of the state that offer the greatest appeal to fourteener seekers. Second only to the Sawatch Range in concentration of the state's tallest peaks, the San Juan Range offers an entire season's worth of mountain climbing in relatively close proximity."[108]

Given these lofty statistics, one might conclude, correctly, that Colorado has the nation's most fourteeners (54 of 91)—which includes 15 in California, 21 in Alaska, and one in Washington—and more than any country in the world. Gerry Roach says the goals and accomplishments associated with Colorado's fourteeners are as numerous as the people who climb them, but some are content just to look at them from afar. Others are excited if they manage to climb one. Many settle for doing just the Class 1 (fairly easy) and Class 2 (somewhat tougher) peaks (see *Colorado's Fourteeners*, p. xix), leaving the others to more adventurous and possibly less intelligent types.[109]

Then there are those who only climb by way of bribes or deception. "Ed Quillen climbed Mt. Elbert in 1979 with Rex Ewing and Allen Best, who kept telling him there was a great Mexican cantina on the top with steaming chili verde and Negra Modelo at $1 a pitcher."[110] While Class 3 routes often involve scrambling, Class 4 is entering the realm of technical climbing, explains Gerry Roach. You are not just using handholds, but have to search for, select, and test them. Your movement is more focused and slower. Many people prefer to rappel down a serious Class 4 pitch rather than down-climb it.[111]

Technical climbing (vs. hiking or scrambling) routes are assigned grades in

---

105 Virginia McConnell Simmons. "Naming the Indian group of the Sawatch Range." *Colorado Central Magazine*: June 2005.

106 Robert Weller. "Fourteeners challenge spirit, legs." *Rocky Mountain News*: 7/14/01.

107 http://www.fourteenerworld.com/.

108 Scott Willoughby. "Colorado's 14ers perfect for summit meetings." *The Denver Post*: 7/6/10.

109 Gerry Roach. *Colorado's Fourteeners*. Golden, CO: Fulcrum Publishing, 1999, p. xxiii.

110 Ed Quillen. "A mountain by any other name would soar as high." *Colorado Central Magazine*: May 1998.

111 Gerry Roach. *Colorado's Fourteeners*. Golden, CO: Fulcrum Publishing, 1999, p. xix.

ascending order of difficulty on an open-ended scale starting at 5.0 and currently topping out at 5.15b.[112] None of Colorado's fourteeners require Class 5 climbing on their standard routes, but the three hardest centennial thirteeners do: Jagged Mountain (13,824 ft.), Teakettle Mountain (13,819 ft.), and Dallas Peak (13,809 ft.).[113]

Each year (notes *MyMidwest* contributor Andrea Bahe), about half a million people, from weekend warriors to scout troops, grab their trekking poles and head to the top of Colorado's fourteeners. They're rewarded, she says, "not only with breathtaking views, but also a journey that brings them face-to-face with the real Colorado: wild rivers, ancient alpine tundra, herds of elk and white-bearded mountain goats and eerie ghost towns."[114]

For some, Bahe adds, summiting a fourteener is a bucket list kind of goal—one that's thrilling but more attainable than, say, Mount Everest. Many of the peaks are clustered along the Front Range, making for easy day trips from Denver. In addition, nearly every fourteener has routes that are called "walk-ups," meaning you can hike a trail from base to summit without climbing gear. But don't let the name fool you, Bahe warns: "Even walk-ups can loop dizzyingly close to cliff edges or require endless scrambling over boulders the size of dishwashers."[115]

Fourteener completer Ricardo Pena adds, "I love the simplicity of an easy hike up a peak in the Sawatch or Tenmile Range [an extension of the Mosquito Range]; the spectacular rock and steep peaks in the Sangre de Cristos; the wilderness character of the San Juans; and the challenging nature of the Elks."[116] Ricardo knows from experience that all fourteeners aren't created equal. Many are exposed, white-fisted, sphincter-loosening rock scrambles. On the other end of the spectrum, the seven peaks listed below (as chronicled by Dave Philipps) are merely steep walk-ups. Anyone willing to sweat and grunt some can reach the top:[117]

Mount Elbert – Round-trip 7 miles, 4,100 feet of elevation gain
Huron Peak – Round-trip 10 miles, 3,700 feet of elevation gain
Mount Princeton – 6 miles, 3,370 feet of gain
Mount Antero – 4.5 miles, 3,400 feet of gain
Mount Sherman – 5 miles, 2,436 feet
Quandary Peak – 4.5 miles, 2,545 feet
Mount Bierstadt – 4.5 miles, 2,400 feet

---

112 Brendan Leonard. "Climb Like Kor." *Trail & Timberline*: Spring 2010, p. 15.

113 http://www.fourteenerworld.com/.

114 Andrea Bahe. "Step It Up." *MyMidwest*: May/June 2010.

115 Andrea Bahe. "Step It Up." *MyMidwest*: May/June 2010.

116 Linda K. Crockett. "The fourteener files." *Trail & Timberline*: Winter 2009, p. 17.

117 Dave Philipps. "Fourteeners 101." *The [Colorado Springs] Gazette*: 3/28/07.

Want to know what or whom we've named some of our state's highest peaks after (again, from Dave Philipps in "Fourteeners 101")?[118]

19 after the peak's physical appearance
19 after dead white men
5 after universities
5 after American Indians
2 after bouts of confusion
2 after peripheral immortals
1 after a political party
1 after a Midwestern state
1 after a crashed space shuttle
1 after a fictitious passage to the center of the Earth

Even though Euro-Americans named (or renamed) most of Colorado's fourteeners, Native Americans were the original Colorado mountaineers. Researchers have found 11,000-year-old chipped stone points high on some peaks, while cliff dwellings like those at Mesa Verde testify to ancient peoples' skill on precipitous terrain. Utes and Arapahoes created numerous trails across high mountain passes, later used by European trappers and explorers, and Native Americans may have built shelters for lookouts on some peaks.[119]

Climbers making the first recorded ascent of Blanca Peak (14,345 ft.) in the Sangre de Cristo Range found a circular excavation that appeared to have been made by Native Americans.[120] Possibly taking their cues from local Indians, miners in Colorado during the mid- and late 1800s often climbed peaks on their days off, and in 1912, the Colorado Mountain Club (CMC) was founded, offering a clearinghouse for climbing information and a fraternity of peak baggers.[121]

Charter members included Enos Mills, whose efforts were influential in establishing Rocky Mountain National Park; Roger Toll, who held the prestigious positions of superintendent at Yellowstone, Rocky Mountain, and Mount Rainier National Parks; and Carl Blaurock, who, along with William Ervin, was the first person to climb all of Colorado's 14,000-foot peaks, or at least the first to climb all of the fourteeners then recognized.[122]

The number of fourteeners has changed multiple times over the years as

---

118 Dave Philipps. "Fourteeners 101." *The* [Colorado Springs] *Gazette*: 3/28/07.
119 Bradford Washington American Mountaineering Museum: Golden, Colorado.
120 Bradford Washington American Mountaineering Museum: Golden, Colorado.
121 Deb Accord. "For The Love Of...Fourteeners/Climbers who scale 54 tallest peaks." *The* [Colorado Springs] *Gazette*: 4/19/01.
122 http://www.cmc.org/about/about_mission.aspx.

surveying techniques steadily improved, and in 1914 the Colorado Mountain Club published its first list of "The High Peaks of Colorado," which included 42 fourteeners.[123] In 1972, 53 Colorado peaks were recognized as fourteeners. Today, 54 mountains are officially deemed to be fourteeners by the CMC. The most recent addition to the list: Ellingwood Point (14,042 ft.).[124]

In 1920, according to the CMC's *Guide to the Colorado Mountains*, two members came up with the idea of keeping lists. Carl Blaurock and William Ervin sat on the summit of Mount Eolus counting the number of fourteeners they had climbed and decided to try and be the first to climb them all. They succeeded in 1923. By 1940, there were seven finishers, which increased to 53 in 1950 and 500 in 1990.[125]

By August 2003, when I completed the 14ers, there were 1,068 finishers. Linda K. Crockett (the CMC's 14er completer list-keeper) says, "To climb all the 14er summits on the list, even by their easiest routes, is a test of endurance and perseverance that often needs to be bolstered with logistical support and a touch of luck."[126] Crockett's name is on the completers list too. It took her 17 years, but she did it "because of the views from the summit."[127]

Linda adds, "In the summer, when other people are enjoying lazy days, barbecues, or relaxing at the beach, the fourteeners crowd is busy with presunrise starts, prodigious drives, and physically grueling days as they make pilgrimages to the summits of Colorado's mountains."[128] The number of people who have summited all 54 now (as of 2008) exceeds 1,200.[129]

"The draw to climb fourteeners has astronomically increased," explains Rodney Ley, assistant director of Colorado State University's Outdoor Adventure Program. "The main overall reason is that there are more people who want to do outdoor adventures. Fourteeners are dramatic, goal-oriented, and simple."[130] But how many mountains (fourteeners and otherwise) can one climb in Colorado?

Trail & Timberline contributor Chris Ruppert (in "Beyond the Fourteeners") says good counts have been made of the 14ers, 13ers, and 12ers, but no authoritative counts of the lower mountains was completed until 2006, when a

---

123 Woody Smith. "Place-name polka: Sorting out the Crestones." *Trail & Timberline*: Fall 2008, p. 34.

124 http://www.fourteenerworld.com/.

125 Deb Accord. "For The Love Of...Fourteeners/Climbers who scale 54 tallest peaks." *The* [Colorado Springs] *Gazette*: 4/19/01.

126 Linda K. Crockett. "The 14er Files." *Trail & Timberline*: Winter 2007-2008, p. 15.

127 Deb Accord. "For The Love Of...Fourteeners/Climbers who scale 54 tallest peaks." *The* [Colorado Springs] *Gazette*: 4/19/01.

128 Linda K. Crockett. "The fourteener files." *Trail & Timberline*: Winter 2009, p. 17.

129 Linda K. Crockett. "The 14er Files." *Trail & Timberline*: Winter 2007-2008, p. 15.

130 Sarah Rawley. "'Tis the season to climb fourteeners." *The Rocky Mountain Collegian*: 9/13/05.

group assembled to make a thorough count. After checking and rechecking, but not claiming to have missed a mountain or two, they concluded there are 4,345 mountains in Colorado. Here's the breakdown by thousand-foot intervals:[131]

14ers: 54
13ers: 584
12ers: 676
11ers: 468
10ers: 530
9ers: 617
8ers: 711
7ers: 455
6ers: 201
5ers: 44
4ers: 6

Although Colorado has over 4,300 mountains by this count, including the most 14ers and the highest average elevation in the United States, it doesn't have America's highest peak (that's Alaska's Mount McKinley, elevation 20,320 feet), nor even the highest mountain in the Lower 48 states (California's Mount Whitney, elevation 14,494 feet), both, of which, top Colorado's king of the hill, Mount Elbert (14,433 feet).[132]

Nonetheless, Colorado has the second, third, and fourth highest peaks in the contiguous 48 states (Mt. Elbert, Mt. Massive, and Mt. Harvard), 54 of the 91 highest peaks in the U.S. (including Alaska), and more than 630 summits over 13,000 feet, making for a maze of wild and mesmerizing mountain terrain one could spend a lifetime exploring and still only scratch the surface.

Scott Willoughby says, "If there is any truth to Polish climber Voytek Kurtyka's proclamation that 'alpinism is the art of suffering,' then Colorado may very well be sculpted in misery. With the highest concentration of tall peaks in the Lower 48, more than a third of the state designated as public land and some of the most active residents in the nation, the local landscape is rife with artistic inspiration of the alpine sort."[133]

Misery, suffering, and alpine inspiration aside, Albert Einstein once said that the most beautiful experience we can have is "the mysterious." It is "the fundamental emotion which stands at the cradle of true art and true science." Einstein felt that, "He to whom this emotion is a stranger, who can no longer pause to

131 Chris Ruppert. "Beyond the Fourteeners." *Trail & Timberline*: Winter 2006-2007, p. 11.
132 Penelope Purdy. "Take mountains to heart." *The Denver Post*: 1/11/02.
133 Scott Willoughby. "Wilderness survival worth a closer look." *The Denver Post*: 9/10/07.

wonder and stand rapt in awe, is as good as dead: his eyes are closed." Einstein was right.

There is undeniably nothing more beautiful and mysterious than that which is found in the wild and natural (the real) world, and in my experience, true beauty and genuine mystery are the provinces of nature and nature alone. What follows are recollections from four years of wandering, exploring, and climbing among 54 fourteeners in the Rocky Mountains of Colorado and other public lands across the country, continent, and world.

"Not all those who wander are lost."

–J.R.R. Tolkien[134]

---

134 *The Fellowship of the Ring*. Houghton Mifflin.

David in Yosemite National Park's Mariposa Grove enroute to Vandenberg Air Force Base, California, for Undergraduate Missile Training: 17 May 1991

DEPARTMENT OF THE AIR FORCE
4315TH COMBAT CREW TRAINING SQUADRON (SAC)
VANDENBERG AIR FORCE BASE, CALIFORNIA 93437-5000

Second Lieutenant David A. Lien
Minuteman II Command Data Buffer
Class 133
Vandenberg AFB CA 93437-5000

3 0 AUG 1991

Dear Lieutenant Lien

Congratulations on your graduation from Minuteman II Command Data Buffer Class 133. Your "mission ready" status represents a significant achievement for both you and the Strategic Air Command.

You have completed a challenging and comprehensive training program. As a "mission ready" graduate, you join a select group of officers on the alert force. I trust you will maintain the high standards this critical duty mandates.

I'm sure you will find your time in missile operations challenging and rewarding. You now have the responsibility to protect the free world. This duty is reserved for only the very best Air Force officers. Be proud you are one of them.

Sincerely

DAVID G. FAAS, Lt Col, USAF
Commander

David's Undergraduate Missile Training graduation certificate: 30 Aug. 1991

David in the 446th Strategic Missile Squadron at
Grand Forks Air Force Base, North Dakota: Fall 1991

David on Kala Pattar (18,370 ft.) in Nepal with Mount Everest in the background: 30 Oct. 2002

David on the north/Tibet side of Mount Everest at 25,200 feet: 19 May 2006

1999: First Fourteeners

Jon Krakauer: "[George] Mallory, whose name is inextricably linked to Everest, was the driving force behind the first three expeditions to the peak. While on a lantern-slide lecture tour of the United States, it was he who so notoriously quipped, 'Because it is there' when an irritating newspaperman demanded to know why he wanted to climb Everest" (*Into Thin Air*, p. 18).[135] Mallory's three expeditions to Mount Everest took place during the 1920s, and his determination to climb the mountain cost him his life.

Edward Abbey: "George H. Leigh-Mallory's asinine rationale for climbing a mountain—'because it is there'—could easily be refuted with a few well-placed hydrogen bombs. But our common sense continues to lag far behind the available technology" (*The Journey Home*, p. 204).[136] Ed Abbey knew from his Colorado fourteener hikes and climbs that there should be, and is, more to climbing than reaching the tops of mountains, but what? "Why?" is the question often asked of climbers, explorers, and other adventurers. Why climb Everest? Why fly solo across the Atlantic? Why go to the moon?

Because climbing and other dangerous endeavors open doors to some of the highest human instincts and emotions—courage, camaraderie, tenacity, fear, sorrow, happiness, and humility. They keep us striving for something, anything beyond mere listless existence. They remind us that life is not meant to be squandered, but lived to the absolute fullest. It takes courage and determination to leave familiar and secure surroundings behind, to embrace the new and unfamiliar, but as conservationist Jodi Broughton said, "There is more security in the adventurous and exciting, for in movement there is life, and in change there is power."[137]

---

135 Jon Krakauer. *Into Thin Air*. New York: Anchor Books, 1997, p. 18.
136 Edward Abbey. *The Journey Home*. New York: Penguin Books, 1977, p. 204.
137 Jodi Broughton. "Journey Ahead Brings Challenges and Change for Conservation." *Northwest Ecosystem News*: Summer 2004, p. 3.

John F. Kennedy understood the importance of change, and why we climb. In one of his many eloquent speeches, President Kennedy said: "Many years ago the great British explorer George Mallory, who was to die on Mount Everest, was asked why did he want to climb it. He said, 'Because it is there.' Well, space is there, and we're going to climb it, and the moon and the planets are there, and new hopes for knowledge and peace are there. And, therefore, as we set sail we ask God's blessing on the most hazardous and dangerous and greatest adventure on which man has ever embarked."

Climbing fourteeners and other peaks in the Rocky Mountains of Colorado or anywhere else falls far short of "the greatest adventure on which man has ever embarked," but summiting Longs Peak in Rocky Mountain National Park was one of my greatest adventures to date, and as with many such "roads less traveled," sometimes not knowing what lies ahead is best; sometimes ignorance is bliss.

Had I read *Into Thin Air* and *The Journey Home* prior to climbing Longs Peak (my first fourteener), I may have reconsidered. Had I known in advance the physical pain and utter exhaustion—the likes of which I had until then never experienced—I'd endure on that 15-mile, 14-hour (round-trip) hike-climb, I may not have attempted it. And had I considered that a man would fall to his death on this day, I may have decided climbing a mountain "because it is there" or for any other reason is not a good enough reason to do it.[138]

Especially a mountain like Longs Peak. *Denver Post* columnist Steve Lipsher (in "Ascent of Danger") notes that with 55 documented fatalities (as of July 2005) since Carrie Welton died of exposure near the Keyhole in 1884, Longs Peak is rivaled only by the Maroon Bells near Aspen as the deadliest mountain in Colorado.[139] Interestingly, the second death on Longs Peak was due to a gunshot wound. On August 28, 1889, Frank Stryker, age 24, died from carrying a loaded pistol in his pocket, which fell out and discharged into his neck as he was scrambling.[140]

Another Longs Peak fatality is memorialized by the highest "building" in the National Park System, the Agnes Vaille Shelter (13,150 ft.) located next to "the Keyhole." It was donated by Agnes's father, F. O. Vaille, and serves as a memorial to Agnes (who died returning from a winter climb in January 1925) and a shelter from the elements for climbers, but it's not one you'd want to be trapped in for very long due to its exposed and lofty locale.[141]

Though many climbers have perished in Colorado's mountains over the years, Vaille's death by exposure was particularly notable due to her wide social,

138 David A. Lien. "First Fourteener: Longs Peak." *Trail & Timberline*: Winter 2006-2007, p. 13.
139 Steve Lipsher. "Ascent of Danger." *The Denver Post*: 7/31/05, p. 8A.
140 http://www.fourteenerworld.com/.
141 http://www.fourteenerworld.com/.

civic, and family circles. Vaille's father installed Denver's first phone system in the 1880s, and made a fortune. In the postmortem examination that followed his daughters' death, it was thought that a shelter may have saved her life, so Mr. Vaille had one built near the Keyhole, where it stands today.[142]

Given the potential dangers associated with climbing peaks like Longs, our wise friend Edward Abbey pondered why we do it, and wrote: "There is something unnatural about walking. Especially walking uphill, which always seems to me not only unnatural but so *unnecessary*. That iron tug of gravitation should be all the reminder we need that in walking uphill we are violating a basic law of nature. Yet we persist in doing it. No one can explain why."[143]

Denver Post columnist Chuck Plunkett (in "Reward Trumps Treachery") can explain. "We went looking for danger. What we found was a route [Long's Keyhole] so spectacular, we wondered how anyone could stay away."[144] Notwithstanding Ed Abbey's feigned disdain for hiking-climbing, he couldn't stay away either and summited (or at least attempted to) a few 14ers and other peaks while frequenting the San Juan Mountains of southwest Colorado.

The time and effort required to reach these lofty places, where relatively few others dare or bother to tread, likely appealed to Ed. On the other hand, had he accompanied me on this Longs Peak excursion, I have no doubt that seeing a dead fellow climber would have fueled much written and verbal condescension from him. But as Italian *überclimber* Reinhold Messner says, "A climber who does not understand that death is a possible outcome of any serious climb is a fool."[145] Ed was no fool, and neither am I.

> "For every potential *Sound of Music* moment, an *Into Thin Air* experience exists. Given enough hours in the outdoors, you're bound to bump into both."
> —Scott Willoughby[146]

AUGUST

Longs Peak was named for Major Stephen H. Long, who led an exploratory expedition of 22 men in 1820 from "Pittsburgh to the Rocky Mountains." The expedition reached the base of the Front Range on June 30, then journeyed south

---

142 Woody Smith. "Agnes Vaille dies in blizzard on Longs Peak." *Trail & Timberline*: Winter 2010, p. 36.
143 Edward Abbey. *The Journey Home*. New York: Penguin Books, 1977, p. 204.
144 Chuck Plunkett. "Reward Trumps Treachery." *The Denver Post*: 8/2/05, p. 6D.
145 Reinhold Messner. *All Fourteen 8,000ers*. Seattle, WA: The Mountaineers, 1999, p. 30.
146 Scott Willoughby. "Wilderness survival worth a closer look." *The Denver Post*: 9/10/07.

where they made the first recorded ascent of Pikes Peak.[147] The Long Expedition officially recorded the location of Longs Peak, even though Major Stephen Long was never closer than 40 miles to the mountain that bears his name.[148]

Writer and historian Francis Parkman first saw Longs Peak in 1846 while riding horseback across Colorado's eastern plains: "Vast piles of clouds were gathered together in the west. They rose to a great height above the horizon, and looking up at them I distinguished one mass darker than the rest... but while the clouds around it were shifting, changing, and dissolving away, it still towered aloft in the midst of them, fixed and immovable. It must, thought I, be the summit of a mountain; and yet its height staggered me."[149] In 1877, the Hayden Survey estimated the elevation of Longs Peak at 14,271 feet, only 12 feet off today's official elevation of 14,259 feet.[150]

Gerry Roach, author of *Colorado's Fourteeners*, says Longs Peak is unquestionably the monarch of the Front Range and northern Colorado. It dominates all within sight of it. Longs is the highest peak in Rocky Mountain National Park and Boulder County. It's also the northernmost fourteener in Colorado and the Rocky Mountains. Its summit attracts thousands of people each year, and it's one of the most popular peaks in the western United States. "The reason for its popularity is obvious," says Gerry. "Longs enraptures all but the most heartless soul."[151]

There are 113 named peaks and at least 60 that exceed 12,000 feet in Rocky Mountain National Park, topping out on Longs Peak.[152] Longs and its lofty "centennial" (i.e., Colorado's 100 highest peaks) neighbor, Mount Meeker, were known as *"Lex Deux Oreilles"* (The Two Ears) to French trappers who eyed them from the eastern plains.[153]

Outside magazine contributor Nick O'Connell's notes that the mountain beat back numerous summit attempts until 1868, when one-armed Civil War veteran Major John Wesley Powell bulled his way to the top. Powell lost his right forearm at Shiloh in 1862 while serving under Ulysses S. Grant, and went on to make the first known descent of the Colorado River through the Grand Canyon.[154] In 1869, Powell embarked (on May 24) on his 900-mile river trip, a legendary exploration of the Green and Colorado rivers, with nine men and four boats. He

147 Stewart Green. "Longs Peak: Highest Mountain in Rocky Mountain National Park." *About.com:* 7/25/10.
148 Deborah Frazier. "A long life rooted in the Rockies." *Rocky Mountain News:* 3/27/04, p. 26A.
149 Frederick R. Rinehart (ed.). *Chronicles of Colorado.* Niwot, CO: Roberts Rinehart Publishers, 1993, p. 14.
150 John Meyer. "These pictures tell great stories." *The Denver Post:* 1/27/09, p. 12C.
151 Gerry Roach. *Colorado's Fourteeners.* Golden, CO: Fulcrum Publishing, 1999, p. 1.
152 Deborah Frazier. "A long life rooted in the Rockies." *Rocky Mountain News:* 3/27/04, p. 26A.
153 C.W. Buchholtz. *Rocky Mountain National Park: A History.* Boulder, CO: University Press of Colorado, p. 33.
154 Nick O'Connell. "Longs Peak." *Outside* online: 6/18/03.

hauled out on August 30, with six men and two boats.[155]

Major Powell became a national hero after completing his first expedition in 1869, and a second followed in 1872, but the river trip itself was merely the highlight of a five-year survey to fill in the West's blank spots. In 1881, Powell became the second director of the U.S. Geological Survey (USGS). One of his first orders commissioned topo maps of the entire United States.[156]

Long's Keyhole route doesn't call for the use of detailed topo maps or technical climbing gear, but it does require scrambling over steeply sloped rocks along stretches so notorious, says Steve Lipsher, they carry descriptive names like: the Ledges, the Trough, the Narrows, and the Homestretch.[157] *Notorious* is the perfect word to describe this climb, which has the look and feel (and dangers) of a much larger mountain. Hence, Long's Keyhole is not a route recommended for your first 14er, but it is one of the most spectacular and breathtaking standard routes on all of Colorado's 14,000-foot peaks.

On August 14, 1999, I started out from the Longs Peak Ranger Station (at 9,400 ft.) before sunup and met a few other summit-seekers below tree line who had already turned around due to high winds and not having the proper clothing. I kept going. It's a six-mile, nontechnical hike up to the Boulderfield, then a 0.3 mile scramble through the Boulderfield to Long's trademark Keyhole (at 13,160 ft.), where I stopped briefly to take in the breathtaking views of Glacier Gorge before traversing along the narrow ledge system (the Ledges) on Long's west face. Not far past the Keyhole, I encountered something unexpected, startling, and tragic.[158]

A man had fallen and his body was visible some 400 feet below, where it had finally come to rest. He wasn't moving, and his wife and friends were huddled together in a small group watching the initial attempts to reach him. Someone was also calling for help on a cell phone. However, the gusting 40 to 50 mph winds that had likely contributed to his fall would also force any rescue team to hike partway up the mountain. I stopped, as all of us going up did, and asked if I could be of assistance in any way, and wondered if the fallen climber was alive. Given the efforts already underway to reach him and mountain rescue, I continued climbing.[159]

As *Coloradoan.com* contributor Miles Blumhardt explains (in "Climbing Longs Peak"), it's a quarter-mile scramble along the Ledges, with drop-offs of 1,000 feet or more. Then it's 600 vertical feet up the Trough, where sometimes you'll

155 Pat Devereux. "A wilderness of rocks: Canoeing the Green River." *Nevada Appeal*: 6/1/07.

156 Backpacker magazine. "Grand Canyon." *Backpacker*: December 2004, p. 76.

157 Steve Lipsher. "Ascent of Danger." *The Denver Post*: 7/31/05, p. 8A.

158 David A. Lien. "First Fourteener: Longs Peak." *Trail & Timberline*: Winter 2006-2007, p. 13.

159 David A. Lien. "First Fourteener: Longs Peak." *Trail & Timberline*: Winter 2006-2007, p. 13.

be traversing through snowfields. At the top of the Trough awaits the Narrows, which is the most exposed section of the hike. Next, the Notch takes you to the beginning of the Homestretch, where you'll need to use cracks in the rock to climb 200 feet to the summit.[160]

Climbing Longs was an experience that reverberated through the rest of my life; the first significant step in a journey that would eventually lead to the mother of all mountains, Mount Everest, in May 2006.[161] "Soaring more than five miles into the atmosphere," explains *Canyon Courier* contributor Stephen Knapp, "Everest is an unforgiving mass of rock and ice littered with the bodies of those who hadn't the skill or judgment or luck to endure its treacheries. Near its summit, there is a third less oxygen than at sea level, making even simple tasks–tying a bootlace or adjusting a crampon–a universe of fatigue."[162]

As much notoriety as 29,035-foot Mount Everest gets for its life-threatening reputation, the world's second-tallest mountain, K2 (28,251 feet)–thrusting skyward out of the Karakoram Range of northern Pakistan–is regarded as the ultimate achievement in mountaineering, for good reason. Four times as deadly as Everest, K2 has claimed the lives of 77 climbers since 1954. In August 2008, 11 climbers died in a single 36-hour period–the worst single-event tragedy in the mountain's history and the second-worst in the long chronicle of mountaineering in the Himalaya and Karakoram ranges.[163]

Despite (or maybe because of) its notoriously high body count, summiting K2 remains a cherished goal for climbers from all over the globe.[164] Before he faced the challenge of K2, American Ed Viesturs (one of the world's premier high-altitude mountaineers) called it "the holy grail of mountaineering."[165] But K2 is way out of my league, and my Mount Everest attempt is a story for another time and one that's already well-chronicled in books by two of my fellow Everest expedition members: *Dead Lucky: Life After Death on Mount Everest*, by Lincoln Hall; and *Everest: Surviving the Death Zone*, by Ronnie Muhl. Both are engaging, fast reads, and all-around excellent Everest stories.

During my far less formidable 5,000-foot elevation gain, 14-hour round-trip (seven up and seven down) Longs Peak climb, I became so tired and exhausted

---

160 Miles Blumhardt. "Climbing Longs Peak" Coloradoan.com: 7/25/10.

161 See Grand Rapids, Minn., Area Library Program presentations by David Lien: "From the Northwoods to Mount Everest," 11/21/06; "Climbing the Seven Summits," 4/25/07; "America's Wildest National Parks," 7/2/09; "Climbing Mount McKinley," 12/29/09); "Climbing Mount Kilimanjaro/Africa's Wildest National Parks," 7/6/10. Copies of each presentation can be purchased for $15 plus $5 shipping from: ictvinc@mchsi.com.

162 Stephen Knapp. "Casualties, commercialism make Evergreen climber rethink extreme mountaineering." *Canyon Courier*: 8/23/06.

163 http://www.amazon.com/gp/product/0767932501/ref=pe_5060_13189370_snp_dp.

164 http://www.amazon.com/gp/product/0767932501/ref=pe_5060_13189370_snp_dp.

165 http://www.amazon.com/gp/product/0767932501/ref=pe_5060_13189370_snp_dp.

ascending the route's steeper Class 3 sections that after each eight or nine steps I had to stop and rest. Moving any faster would have resulted in getting sick or possibly blacking out (a "universe of fatigue") due to the combined affects of altitude and physical exertion pushing me nearly to utter exhaustion, not unlike what happens to many climbers on more dangerous and deadly mountains, like Mount Everest and K2.

While Longs is no Everest, it was my first 14er and close enough for me at the time. Jon Krakauer says, about climbing Everest, "Every four or five steps I'd have to stop, lean against the rope, and suck desperately at the thin, bitter air, searing my lungs in the process. I reached the top of the serac... and flopped breathless onto its flat summit, my heart pounding like a jackhammer."[166] Jon knows from experience, as do I after attempting Mount Everest, that few things wreak more havoc on your biological hardware than altitude.

I'm not just talking about those who claw their way up the Himalaya's 8,000-meter (26,000-foot-plus) peaks. Heights as modest as 4,000 feet above sea level can affect physical performance and sometime result in death.[167] Altitude sickness is caused by a relative lack of oxygen; the medical term is "hypoxia."[168] High Altitude Pulmonary Edema (HAPE) is a severe type of altitude sickness once thought to affect only 2 to 4 percent of climbers above 8,000 feet. However, a study published in the medical journal *Lancet* found that 60 percent of climbers on the summit of a moderate, nontechnical 14,957-foot peak in the Alps exhibited signs of subclinical HAPE and 15 percent had clinical HAPE.[169]

The highest year-round town in the world is La Rinconada, Peru (16,730 ft.). There are no permanent settlements above this height anywhere on earth because no one can adjust to the altitude year-round. But acute mountain sickness (AMS) can strike as low as 6,300 feet, so climbers, skiers, and even backpackers are at risk—up to 42 percent of visitors to Colorado fall prey to AMS, says Peter Hackett, MD, a Telluride-based altitude specialist. Another oftentimes deadly form of AMS, High Altitude Cerebral Edema (HACE), occurs when leaking fluid causes the brain to swell, whereas HAPE is caused by fluid seeping into the lungs.[170]

A hunter from coastal Washington state flew from sea level to mile-high Denver, then drove to Trapper's Lake on the Western Slope and backpacked to an altitude of 12,500 feet in the 235,400-acre Flat Tops Wilderness Area—all within the span of 18 hours. He spent all five days of the hunt on his back in

---

166 Jon Krakauer. *Into Thin Air*. New York: Anchor Books, 1997, p. 84.

167 Nick Heil. "The Outer Limits: How to handle nature's harshest stuff." *Outside* online: 7/12/03.

168 Highpointers Club. "#24: The Dreaded Diseases of Altitude." *Apex to Zenith*: 4th Quarter 2008, p. 28.

169 Dina Mishev. "Study: Altitude Sickness 'Prevalent'" *Outside* online: 2/5/02.

170 National Geographic. "Elevation Gains." *National Geographic Adventure*: October 2007, p. 44.

agony suffering from altitude sickness.[171]

On July 4, 2005, a 17-year-old from Denver died from altitude sickness after complaining of a headache and cough. He'd been at 10,000 feet in Colorado's 73,000-acre Rawah Wilderness.[172] Federal Aviation Administration (FAA) regulations say that at altitudes between 12,000 and 14,000 feet, pilots need supplemental oxygen to perform adequately if exposed for more than 30 minutes.[173]

Fourteenerworld.com contributor Paul Nesbit describes the stages of AMS: "Mountain sickness: There are three stages. The first one, a person is afraid he is going to die. The second one, he gets so bad that he does not care if he dies, and the third one, he is afraid that he is not going to die."[174] One source claims that for every thousand feet increase in altitude, the human body's work capacity is reduced by 3 percent.[175]

Jim Whittaker, the first American to climb Mount Everest, says the hardest thing was the altitude. "Even with the oxygen tanks, we were just sucking air," he explains. "Put a pillow on your face, run around the block, and try to suck oxygen through that pillow. It will give you an idea."[176] Whittaker summited Everest on May 1, 1963, with Sherpa Nawang Gombu (a nephew of Tenzing Norgay).[177]

For someone like me, who (as of 1999) had never been on a 14er and didn't do anything particularly physically challenging to prepare for the climb, breathing while on the upper reaches of Longs Peak was just about as difficult, but as mountaineer Richard F. Fleck wrote, I "kept on plodding upward as though heaven itself were... [my] destination."[178] After reaching the summit and basically collapsing, then resting for 20 minutes, I started the equally difficult descent. Nearly whipped and only halfway through the climb, I didn't even bother taking in the scenic vistas from Long's unusually flat and spacious football-field-sized summit.

Like Jon Krakauer said about being on the summit of Everest, "Now that I was finally here... I just couldn't summon the energy to care."[179] I was nearly spent, lying flat on my back, and it took all the physical and mental effort I could muster just to stand up and start moving again. Descending slowly down the Homestretch, then across the Narrows and through the Trough, I made it safely back to the Ledges, where the aforementioned climber had fallen several hours

171 Cortez Journal. "Hunter should adjust to high elevations." 2010 Hunting Guide: September 2010, p. 14.

172 Kelly Bastone. "America's Most Dangerous Hikes." Backpacker: October 2008, p. 65.

173 Bibb Underwood. "The First Fourteener." Summit Daily News: 8/24/03.

174 http://www.fourteenerworld.com/.

175 Rick Hartman. "#23 Acclimatization." Apex to Zenith: 3rd Quarter, p. 24.

176 Michael Shnayerson. "Jim Whittaker, Back on Earth." National Geographic Adventure: May 2003, p. 60.

177 http://en.wikipedia.org/wiki/Jim_Whittaker.

178 Richard F. Fleck. "Major Powell's Ascent of Longs Peak: A Fictionalized Account." The Climbing Art Magazine #34, p. 57.

179 Quotes From Everest: www.mnteverest.net.

earlier, and stopped to watch a helicopter recover his body from its resting place on a ledge 400 feet below.[180]

In 1999, four climbing-related fatalities occurred on Longs. In 2000, three climbers died. Although most deaths don't occur on the Keyhole route, Long's fatality statistics are sobering. As of July 2005, falls had killed 33 people, including seven on the Keyhole route. In addition, sudden storms above timberline can bring instant death by lightning, which had happened three times. Five deaths were attributed to heart attacks, and the remainder to a variety of causes. Park rangers conduct an average of 60 rescue missions on the mountain each year.[181]

The peak saw no deaths from 2005 to 2008, and of three deaths in 2009 two were from heart attacks. Even though climbing Longs typically doesn't require ropes, it does have high-risk maneuvers and difficult scrambling. "Fourteener climbing is a strange duck because you are in terrain where it's not practical to use a rope, but you do expose yourself to a lot of fall potential, where if you trip, you die," said Lou Dawson, an expert in high-altitude recreation, guidebook author, and the first to ski all of the state's fourteeners.[182]

When the Keyhole route is snow- and ice-free, it's rated Class 3 with only a few places where exposed scrambling or precarious handholds are encountered, but winter weather transforms it into an "Alpine Ice 1" route, meaning special gear is needed to safely navigate the narrowest ledges and steepest stretches. The first known winter climb of Longs Peak was completed by Enos Mills in February 1903.[183] Enos, a naturalist, climber, and "father" of Rocky Mountain National Park, led many groups to the summit of Longs from his Longs Peak Inn. He climbed the mountain 297 times, including 32 ascents just in August 1906.[184]

"Climbing a high peak occasionally," wrote Mills, "will not only postpone death but will give continuous intensity to the joy of living."[185] You might (or might not) also be interested to know that dogs were allowed on Longs Peak prior to the establishment of Rocky Mountain National Park, on January 26, 1915. In 1906, when Victoria Broughm was staying at the Longs Peak Inn, she expressed an interest in climbing Longs Peak solo, so Enos sent his dog "Scotch" along with her. According to Mills, "Scotch knew the trail well and would... lead her the right way, providing she lost the trail."[186]

Even though the Keyhole is technically a nontechnical climb during most of the summer and early fall, fit for both dogs and man (or woman), many have

180 David A. Lien. "First Fourteener: Longs Peak." *Trail & Timberline*: Winter 2006-2007, p. 13.
181 Steve Lipsher. "Ascent of Danger." *The Denver Post*: 7/31/05, p. 8A.
182 Jason Blevins. "Deadly Ascents." *The Denver Post*: 10/8/10, p. 18A.
183 http://www.fourteenerworld.com/.
184 Stewart Green. "Longs Peak: Highest Mountain in Rocky Mountain National Park." *About.com*: 7/25/10.
185 Stewart Green. "Longs Peak: Highest Mountain in Rocky Mountain National Park." *About.com*: 7/25/10.
186 http://www.fourteenerworld.com/

died in these "nontechnical" climbing conditions, including a guy who rolled the dice with his life and succumbed to exposure on September 4, 2004. "He was totally dressed in cotton and was caught in a storm," said park ranger Jim Detterline, who found the man's body leaning against a route marker on Long's summit.[187]

"If he had turned around, he would have been cold and wet but he would have been all right," said Detterline.[188] Instead, he's dead. According to Gerry Roach, somehow Longs' popularity makes people feel safer, but the opposite is the case: "Any route on Longs is a serious undertaking. The Keyhole route is a long, arduous ascent on a high, real mountain. The route's difficulty increases dramatically when conditions are bad," and conditions can go from bad to nightmarish fast.[189] Two different weather stations on the summit have been destroyed by winds clocked at more than 220 mph.[190]

In his book *Mountains of The Great Blue Dream*, Robert Reid writes about the intermingling of climbing and death: "In the curious playgrounds of their sport, mountaineers learn what primitive people knew instinctively–that mountains are the abode of the dead, and that to travel in the high country is not simply to risk death but to risk understanding it." In *This Game of Ghosts*, Joe Simpson says the reason for death being so essential to mountaineers is that it enables us to see life for what it truly is.[191]

Joe and Robert both know that climbing prepares one for death, leads one toward the edge of another world into which we can look without fear. In the urban world, our greatest and deepest anxiety is the fear of death, but in the natural world of the mountains it is possible, Reid says, to overcome this fear. Far from being separate from life, death is really a smooth continuation from life, and for this reason, mountaineers can move easily to the edges of each world.[192]

> "There has been so much to be said for being
> in the good company of the dead."
> –George L. Mallory

This Longs Peak climb was my first close encounter with death in the mountains, but death and I had crossed paths before. My parents died during July

---

187 Don Hopey. "Window closing on safe ascent of peak in the Rockies." *The Boston Globe*: 9/20/05.

188 Don Hopey. "Window closing on safe ascent of peak in the Rockies." *The Boston Globe*: 9/20/05.

189 Gerry Roach. *Colorado's Fourteeners*. Golden, CO: Fulcrum Publishing, 1999, p. 1-3.

190 http://www.fourteenerworld.com/.

191 Joe Simpson. *This Game of Ghosts*. London: Random House, 1993, p. 232.

192 Joe Simpson. *This Game of Ghosts*. London: Random House, 1993, p. 232.

1975, when I was seven years old. Their early deaths showed my sister and me the frail and precarious nature of life at an age when such thoughts should never cross a child's mind, but it also abruptly and traumatically began the process of teaching me how to fully embrace life, so as not to fear death when it inevitably comes my way.

While descending Longs Peak, I contemplated the day's tragic events, accompanied by one other equally tired and sluggish climber. We were nearly the last ones off the mountain and made it back to the relative safety of the trees just as an afternoon thunderstorm was gathering steam, but encountering herds of elk loitering in the high basins above tree line was ample reward for our tortoise-like pace. At dusk, finally nearing the trailhead, we stopped and chatted with two climbers from the Midwest who had driven straight through from Ohio to climb Longs.

They were backpacking up to the Boulderfield (a talus plain below the Keyhole) to camp for the night and climb in the morning. A sensible way to do Longs for most people. Incidentally, the highest designated campsites in the National Park System are found in the 12,760-foot Boulderfield. From the Longs Peak trailhead, it's a six-mile hike to the nine rock- and boulder-strewn tent sites located there.[193]

After taking a look at me, absolutely spent and barely capable of walking or talking, the Ohio climbers surely must have realized they were doing Longs the right way, and were amazed we had done the 15-mile round-trip in one shot. Me too. We talked for a few minutes and found out they were heading for Mount Elbert next, which was also destined to be my next 14er, a thankfully much easier hike-climb than Longs.

I started climbing Longs Peak before sunup and returned after sundown. Physically, at that time, it was one of the hardest days of my life, mostly because I wasn't in shape for such a long, rigorous 14er ascent. It was also extremely windy and cold on the more exposed sections of the route, but without a cloud in the sky. Because I was cautious and prepared, the weather was not enough to turn me back or take my life, as it did others this day. I completed the climb at a snail's pace and finished utterly exhausted, and may have even been last off the mountain, but I was alive, still on my feet, satisfied with the results, and ready for more.

Mount Elbert (14,433 ft.) is the highest peak in Colorado and in the entire Rocky Mountains, and the second-highest in the Lower 48 states, after California's Mount Whitney (14,494 ft.). Elbert is located in central Colorado about 140

---

193 Backpacker. "Where Can I Find It?" *Backpacker:* August 2008, p. 37.

miles west of Denver, 50 miles south of Vail, and 40 miles east of Aspen.[194] The peak sits in plain view 12 miles southwest of Leadville–the highest incorporated city in the United States (10,430 feet)–but many people mistake her for a thirteener. "With a more intense inspection," says Gerry Roach, "you can see Elbert for the monarch it is."[195]

The only U.S. World War II Army division trained to fight in the mountains in winter conditions was the 10th Mountain Division, which trained at Camp Hale, Colorado, near Leadville. The volunteers learned winter and mountain survival techniques, rock climbing, and combat skills during the winter of 1943-44.[196] According to *Backpacker* magazine, Leadville was also "once the West's wildest and richest boomtown... famous for flamboyant characters like Doc Holiday, Kit Carson, and Buffalo Bill Cody. Today, this quaint town at the headwaters of the Arkansas River is a bustling gateway to some of Colorado's sweetest peaks," like Mount Elbert.[197]

Located in the 14er-strewn Sawatch Range, Mount Elbert was named after Samuel Elbert, who was the territorial governor of Colorado in 1873. The first recorded Mount Elbert ascent was completed by H. W. Stuckle of the Hayden Survey in 1874. *Backpacker* says, "Elbert climbers typically ply the 5.5-mile (one-way) Black Cloud Trail, which passes through a gorgeous stand of aspens before turning northeast onto the peak's expansive flank. There's no exposure or technical terrain to slow you down–though thin air and four false summits lengthen the day."[198]

As a bonus, you can also summit South Elbert (14,134 ft.) on this route, although it's not an official fourteener. Summitpost.org adds, "There are five main routes up Mount Elbert. The two most popular, and easiest to climb, are the South Mount Elbert Trail and the North Mount Elbert Trail, both well-trodden Class 1 routes."[199] I used the nine-mile (round-trip), 4,400 foot elevation gain North Elbert Trail. The day started out partly cloudy and the summit socked-in early, but misty precipitation (no lightning or thunder to speak of) was the only weather-related challenge. One other climber and I had the mountain to ourselves, being the only glutton-for-punishment peak baggers willing to brave the morning's fog and light drizzle to reach the summit on this dreary day.

While standing on the summit at that particular instant, we were higher than everyone else in the Lower 48 states, except perhaps for a few climbers

---

194 http://www.summitpost.org/mountain/rock/150325/mount-elbert.html.

195 Gerry Roach. *Colorado's Fourteeners.* Golden, CO: Fulcrum Publishing, 1999, p. 93.

196 http://www.fourteenerworld.com/.

197 Backpacker magazine. "Mt. Elbert, Colorado." *Backpacker*: December 2004.

198 Backpacker magazine. "Mt. Elbert, Colorado." *Backpacker*: December 2004.

199 http://www.summitpost.org/show/mountain_link.pl/mountain_id/153.

who might have been on the summit of Mount Whitney (61 feet higher) at that same moment. As most Colorado hockey fans know, the Stanley Cup has been hoisted high many times, and once as high as the top of Mount Elbert. After the Colorado Avalanche won the Stanley Cup in 2001, a group led by Mark Waggoner (the Avalanche's vice president of finance) carried the Cup to Elbert's summit. Since the trophy weighs 35 pounds and stands almost 3 feet high, it was no easy feat, but Waggoner is an avid climber and this was his third time on Elbert.[200]

Such shenanigans have increasingly put Mount Elbert and Colorado's other 14ers in the national spotlight. Partly as a result, being alone (or nearly so) on Mount Elbert (Stanley Cup in tow or not) and many other 14ers is an increasingly unusual experience. According to *Trail & Timberline* contributor Brendan Leonard (in "Fourteener's Initiative saves peaks from being 'loved to death'"), during the past decade, the number of climbers attempting to bag Colorado's fourteeners has increased by an estimated 300 percent, and the numbers are growing 10 percent each year. In total, the fourteeners are visited by nearly 500,000 people yearly (as of 2004).[201]

Anya Byers (recreational planning coordinator for the CMC) says, "Those numbers are particularly skewed, as 80 to 90 percent of visitors stick to the easily accessible, nontechnical peaks located in the Front Range, Mosquito Range, and Sawatch Range, of which there are about 20. Worse still, all of this climbing gets squeezed into the short alpine summer months of July, August, and early September. It's no wonder the parking lots fill early. A hiker on a popular fourteener in summer is guaranteed to be sharing the trail with at least 200 to 400 other eager, peak-bound strangers."[202]

"There's a reason people are out here in the mountains," blind climber and Colorado resident Erik Weihenmayer says. "It's because open spaces add to our quality of life."[203] Weihenmayer is the author of *The Adversity Advantage* and *Touch the Top of the World*, and he's famous for climbing Mount Everest and the Seven Summits. In fact, Erik is the only blind person in history to reach the summit of Mount Everest, a feat he accomplished in 2001. In addition, he's scaled the 3,000-foot-high face of El Capitan in Yosemite National Park, skied down the tallest peak in Europe, and guided blind Tibetan teenagers to 21,500 feet on the north side of Mount Everest.[204]

---

200 CBC Sports. "Avs take Stanley Cup mountain climbing." *CBC Sports*: 8/22/01.

201 Brendan Leonard. "Fourteener's Initiative saves peaks from being 'loved to death.'" *Trail & Timberline*: October-November 2004, p. 24.

202 Anya Byers. "Tread Lightly: Saving Our Fourteeners One Step at a Time." *Trail & Timberline*: Spring 2010, p. 12.

203 Joshua Zaffos. "Back on Top." *Land & People*: Spring/Summer 2008.

204 Brenda Porter. "Mountain Fest slated for October 18." *Trail & Timberline*: Fall 2008, p. 18.

In 1997, Weihenmayer moved from Phoenix to Colorado after sprawl and development cut off access to Pinnacle Peak and other climbing spots around his former hometown. In Colorado, access is becoming an issue too, as is the sheer number of climbers trudging toward the tops of 14,000-foot peaks. About 60,000 people climbed a fourteener in 1984, according to the Colorado Fourteeners Initiative (CFI). By 2004, the number swelled to half a million, and the CFI predicts there will be one million yearly fourteener climbs by 2011.[205]

While some of the more remote peaks remain pristine, increasing recreational use has noticeably and negatively impacted many of the more accessible mountains and their alpine basins. Gladly, the CFI was founded in 1994 to "protect and preserve the natural integrity of Colorado's 54 14,000-foot peaks and the quality of recreational experiences they provide."[206] The CFI, as its Web site says, is a "partnership for preservation."

CFI has partnered with the U.S. Forest Service to tackle damage on 37 of the 14ers—those in the most critical condition during the Forest Service's initial fourteener impact study in 1994. Minimizing human impact is especially important on peaks found in wilderness areas, where 32 of the fourteeners are located, because, according to the 1964 Wilderness Act, these areas should be "affected primarily by the forces of nature, with the imprint of man's work substantially unnoticeable."[207]

The CFI's executive director, James Ashby, says: "Some of the fourteeners are seeing over 400 hikers per day during the summer. It's our hope that anyone who is concerned about the impacts of that number of people on the fragile alpine ecosystem will give back to the peaks by volunteering with CFI."[208] One CFI project involved hauling in more than 400 tons of rock to fill a quarter-mile-long, four-foot-deep gully caused by erosion on Humboldt Peak (14,064 ft.), according to *Time* magazine.[209]

CFI also facilitates Partnership Projects, where volunteers spend up to a week rebuilding trails and restoring natural habitats. "Most of the fourteener trails aren't in sustainable locations," Field Program Manager Greg Seabloom says. "So we harden them with rock steps, staircases and walls. If needed, we reroute trails and then restore and re-vegetate the old ones."[210] In addition, CFI spon-

---

205 Joshua Zaffos. "Back on Top." *Land & People*: Spring/Summer 2008.

206 Brendan Leonard. "Fourteener's Initiative saves peaks from being 'loved to death.'" *Trail &Timberline*: October-November 2004, p. 25.

207 Brendan Leonard. "Fourteener's Initiative saves peaks from being 'loved to death.'" *Trail &Timberline*: October-November 2004, p. 25.

208 Jim Gehres. "Colorado Fourteener Initiative celebrates fourteen years of volunteerism on the 14ers." *Trail & Timberline*: Spring 2008, p. 14.

209 http://www.fourteenerworld.com/.

210 Andrea Bahe. "Step It Up." *MyMidwest*: May/June 2010.

sors volunteer weekends to bring in extra help.

Volunteers bring their own camping gear and CFI provides the food and entertainment. The group operates from a small office in Golden, Colorado, and each summer sends out field crews to base camps near peaks needing remediation work. During the summer of 2006, for example, operations were underway on Pyramid Peak, Mount Evans, and Mount Massive.[211]

As of February 2007, CFI had conducted trail restoration and delineation work on 17 fourteeners, including: Mount Elbert, Mount Belford, Mount Oxford, La Plata Peak, Humboldt Peak, Huron Peak, Mount Harvard, Grays and Torreys peaks, Missouri Mountain, Mount Bierstadt, Quandary Peak, Capitol Peak, Tabeguache Peak, Mount Sneffels, Wetterhorn Peak, and the upper portion of Mount Evans.[212] It was a welcome sight to see the results of CFI's hard work on some of the 14ers I climbed next.

September

After completing Longs Peak and Mount Elbert during August, between September 4 and November 20 I climbed six more fourteeners: Grays Peak, Torreys Peak, La Plata Peak, Mount Massive, Humboldt Peak, and Mount Bierstadt. The popular and relatively easy Grays and Torreys combo was up first (more on them in a minute), and the following weekend, I was turned back on La Plata Peak due to questionable weather. La Plata was to be my fifth 14er, and like a young hunter overly focused on the kill, I was intent on adding another mountain to my growing list of summits, but commonsense prevailed and turned me around well short of the summit, something that would happen multiple times over the years.

According to Summitpost.org, a Hayden Survey team climbed La Plata Peak (14,336 ft.) on July 26, 1873, naming it "La Plata," which means "silver" in Spanish. The Hayden Survey, as previously mentioned, was a contracted effort by the U.S. government to triangulate locales and map the American West. The survey team members found out for themselves that La Plata Peak has steep surrounding slopes and gentle, broad ridges leading to an expansive summit plateau, which is Colorado's 5th highest.[213]

Mountain weather is subject to rapid change, especially during the prime summer months. Summitpost.org reminds climbers about mountains brewing their own localized weather systems and emphasizes that forecasts from nearby towns often have little to do with actual conditions. During the summer and early <u>fall, you can</u> generally expect clear to partly cloudy skies in the morning, fol-

211 Carol Kauder. "Trail work in backcountry requires lots of sweat equity." *Daily Camera:* 8/18/06.
212 Colorado Fourteeners Initiative (CFI). Organizational website (accessed February 11, 2007): *www.fourteeners.org.*
213 http://www.summitpost.org/show/mountain_link.pl/mountain_id/203.

lowed by fast-developing and -dissipating thunderstorms in the afternoon. There is no substitute for getting an early start and putting as much of the mountain behind you as early as possible.[214]

A strange (to me) but not uncommon combination of snow and thunder convinced me to turn around on La Plata Peak. I was learning from experience that caution and commonsense are the trademarks of almost all successful climbers, like six-time American Mount Everest summiter Ed Viesturs. "A lot of people are willing to continue on, risk their lives," he says. "I'm not. We probably could have made it to the top, but with the conditions and our abilities, we weren't sure we could make it down. And that's the critical factor. Getting up is optional. Getting down is mandatory. It's gotta be a round trip."[215]

No summit is worth knowingly risking your life over, and if it is, you're climbing for the wrong reasons. Everest veteran Conrad Anker agrees. "Being a good climber also means that you have good judgment. Climbing is a cerebral sport and not just about how strong you are."[216] Viesturs adds, "If I didn't enjoy the process, I couldn't have done summit after summit. There's got to be more to it. You've got to kind of revel in the whole process... the travel, the planning, the camaraderie, being in a place where you all love being. If the mountain said, 'You're not going to the summit this year,' at least we can say we had a good time trying."[217]

Minimizing risk, reveling in the process, and enjoying the outdoors combined with using good judgment and respecting the weather and its inherent unpredictability are critical in the mountains. If you don't know that, you don't know much of anything about wildlands (much less mountains) and are better off staying at home, out of the wilderness that demands our utmost attention, respect, and defense. Like one of my fellow climbers said, "The great mountaineers are the ones who know when to come back another day, not the ones that go out at any cost and end up dead."

Ed Viesturs probably knows this mountaineering truism better than anyone. Of all the world's highest, most treacherous peaks, only 14 summits reach above 8,000 meters (26,200 feet-plus) in elevation. These heralded "Eight-Thousanders" include the likes of Everest, K2, Makalu, Annapurna, and 10 more stratosphere-piercing summits. Ed has stood atop each one: "It took 18 years and 29 individual expeditions," he says.[218]

Ed was the first American and only the twelfth person ever to reach the sum-

214 http://www.summitpost.org/show/mountain_link.pl/mountain_id/341.
215 Craig Vetter. "Ed Viesturs." *Outside*: December 2000.
216 Mark Synnott. "Lofty Goals." *Hemispheres*: November 2008, p. 67.
217 Joe Robinson. "He beat the beast." *Los Angeles Times*: 5/24/05, p. F6.
218 Stephen Regenold. "Best mountains to climb in your lifetime." *Forbes Traveler.com*: 12/14/06.

mits of all fourteen 8,000 meter peaks. In 1989, Viesturs stood atop his first 8,000er, on 28,169-foot Kanchenjunga in Nepal. In 1990, he climbed 29,035-foot Mount Everest for the first of six times and has been on the summits of 8,000 meter peaks more than 20 times (as of 2003)—more times than any man. Jim Whittaker, the first American to climb Everest, calls Viesturs the best American mountain climber ever.[219]

David Breashears—a mountaineer, filmmaker, and close friend of Viesturs—joined Ed on two Mount Everest expeditions. David says, "What sets Ed apart is his tremendous ambition to be the first American to climb all [the] 8,000 meter peaks. It's his personal goal. He would be doing it with or without sponsors and with or without publicity." Most importantly, David adds, "Ed has managed to have a life in the high mountains, and he still has not only his life, but all his fingers and toes."[220]

In April 2010, a South Korean mountaineer became the first woman to scale the Eight-Thousanders, crawling on all fours (reports the Associated Press) as she reached her last summit. Oh Eun-sun, 44, summited Annapurna 13 hours after she left high camp. Annapurna was the last of the 14 peaks above 26,247 feet (8,000 meters) she needed to set the mark.[221] It's also one of the world's deadliest peaks, with a fatality rate exceeding 50 percent.[222] Oh reached the summit—26,545 feet (8,091 meters)—13 years after scaling her first Himalayan mountain, Gasherbrum II, in 1997. She climbed Mount Everest in 2004.[223]

Stephen Venables was the first Briton to climb Mount Everest without oxygen. His tale of surviving a night near the summit of the highest place on the planet has gone down in mountaineering folklore as a tale of amazing fortitude. The Times contributor Simon Crompton (in "Mountaineer Stephen Venables on staying alive") describes what happened: "It was in 1988 that Venables undertook his famous ascent of Everest... Whipping off his left shoe and sock, he shows me the horny stumps where his toes should be. Frostbite got them on Everest. 'It's quite common for mountaineers to lose them,' he says dismissively... Only 20 others had achieved the climb without oxygen and four died on the way down."[224]

Ed Viesturs, Oh Eun-sun, Stephen Venables, and those like them are among tens of thousands of sea level-averse souls afflicted by a compulsion for high altitude adventure. And today, growing numbers of nonprofessional climbers

219 Jim Cour. "Renowned American mountaineer shoots for milestone climb." Duluth News Tribune: 10/18/03.

220 Jim Cour. "Renowned American mountaineer shoots for milestone climb." Duluth News Tribune: 10/18/03.

221 Binaj Gurubacharya. "S. Korean climbs into history." Associated Press: 4/30/10.

222 Lori Spaulding. "Experiences on Annapurna remind climber of the unforgiving nature of the mountains." PikesPique: May 2009, p. 1.

223 Binaj Gurubacharya. "S. Korean climbs into history." Associated Press: 4/30/10.

224 Simon Crompton. "Mountaineer Stephen Venables on staying alive." The Times: 12/27/08.

and adventurers are attempting to reach not just the summits of peaks like Mount Everest, but the highest peaks on each continent (dubbed the "Seven Summits"): Mount Kilimanjaro (19,340 ft.) in Africa, Aconcagua (22,835 ft.) in Argentina, Mount McKinley (20,320 ft.) in Alaska, Mount Vinson (16,067 ft.) in Antarctica, Mount Elbrus (18,510 ft.) in Russia, Mount Everest (29,035 ft.) in Asia, and Mount Kosciuszko (7,310 ft.) in Australia.

Some also insist on climbing Carstensz Pyramid, which reaches 16,023 feet on the island of New Guinea, the highest peak in "Oceania" (sometimes called Australasia).[225] The Java Trench, which cuts through the heart of New Guinea, is the geological border that separates Asia from Australia. Carstensz Pyramid lies to the north of the mountain chain that marks the Java Trench, on the Philippine plate of Asia. Thus, it is not on the same tectonic plate as Australia and not one of the Seven Summits, but the debate continues.[226]

For many mountaineers, these peaks represent the Holy Grail. The book *Seven Summits* by Dick Bass and Frank Wells (along with Rick Ridgeway), the men who invented the summits idea, starts many adventurers down the path to climbing these far-flung monsters.[227] Beyond that, some like Ed Viesturs seek the ultimate in alpine summitry—scaling the world's 14 highest mountains. In "Serial Summit Disorder," *Los Angeles Times* contributor Jenny Dubin calls it a habit bordering on "obsessive hand-washing."[228]

"You've got to have a hard-core mental attitude," legendary American climber Fred Becky said. "You've got to have the right mantra. You've got to have dedication, a sense of security, safety and sensitivity with your partners, and a good sense of balance. It's a combination of many, many things. You need to have the capability or desire to accept a certain amount of risk. A lot of it is... spiritual, not a religious type, but you have to have an affinity with the outdoors... You're putting yourself on the line. Man used to put himself on the line all the time. Nowadays we're protected by the police, fire, everything. There's not much adventure left. Unless you look for it."[229]

But "looking for it" is a dangerous and oftentimes deadly pursuit, one that ultimately demands obsessive behavior, because being repeatedly exposed to the best and worst of Mother Nature's unpredictable temperament can change your life in profound ways—and take it away in an instant. Renowned long-distance backpacker Colin Fletcher knew the dangers of wilderness travel well: "I needed

---

225 Shaun Bishop. "Climber well on way to conquer seven summits." *Burlingame Daily News*: 7/29/07.

226 Ronnie Muhl. *Everest: Surviving the Death Zone.* Inspiration at Work Publishing: Cape Town, South Africa, 2008, p. 59.

227 Rick Ridgeway, Frank Wells and Dick Bass. *Seven Summits*. New York: Warner Books, 1986.

228 Jenny Dubin. "Serial Summit Disorder." *Los Angeles Times*: 10/5/04, p. F4.

229 Michael Brick. "At 85, More Peaks to Conquer and Adventures to Seek." *The New York Times*: 12/16/08.

something to pare the fat off my soul, to scare the shit out of me, to make me grateful, again, for being alive. And I knew... there is nothing like a wilderness journey for rekindling the fires of life."[230]

In *Chronicling The West*, Michael Frome says forces like blizzard, cold, drought, earthquake, fire, flood, heat, hurricane, storm, volcano, and wind are beneficial as well as inevitable influences on the planet. In Michael's words, "They are the architects that shape and reshape the land into landscapes and that continually recast form and function of all the creatures, whether plants or animals, growing upon the land and in the water. They are nature's art and poetry and dance and music, the genuine originals that spark creative inspiration in the human soul."[231]

They're also potential killers of hikers, climbers, hunters, mountaineers, and other outdoorsmen and women, but I'd rather take my chances in the mountains and risk dying from an unexpected fall or lightning strike, then fade away in some retched nursing home, especially after having not truly lived, like most people. As A. Sachs said, "Death is more universal than life; everyone dies but not everyone lives."

Another wise adventurer-philosopher said, "Life is not a journey to the grave with the intention of arriving safely in a pretty and well-preserved body, but rather to skid in broadside, thoroughly used up, totally worn out, and loudly proclaiming–WOW–What a Ride!" Frequent hiking, backpacking, climbing, and hunting almost guarantee such a ride. "Some people try to turn back the odometers. Not me! I want people to know why I look this way. I've traveled a long way and some of the roads weren't paved," adds another unknown adventurer.

One of the world's greatest mountaineers and adventurers, Reinhold Messner, knows Mother Nature's wild and indiscriminate temperament better than most. "Mountains are not fair or unfair," he says, "they are just dangerous." In *Stones of Silence*, George Schaller adds: "Mountains are not chivalrous; one forgets their violence. Indifferently they lash those who venture among them with snow, rock, wind, cold."[232]

Instinctively knowing the risks of wilderness and mountain travel from years of hunting, trapping, hiking, camping, and canoeing in the wilds of northern Minnesota, and now facing a possible storm on Colorado's La Plata Peak, I opted to retreat, but promptly returned the following weekend. Blazing a trail through clear blue skies, glistening snowfields, and snow-covered boulders, I reached the summit before anyone else that morning. It was a picture-perfect day, as are most days on any mountain in anyplace wild. Like Ed Viesturs said,

230 Backpacker magazine. "Colin Fletcher, March 1922-June 2007." *Backpacker*: June 2007.

231 Michael Frome. *Chronicling The West*. Seattle, Washington: The Mountaineers, 1996, p. 195.

232 Greg Mortenson and David Oliver Relin. *Three Cups of Tea*. New York: Penguin Books, 2006, p. 108.

"I've never had a bad day in the mountains."[233]

Considering that mountaintops have traditionally been a place of refuge for prophets, particularly as they seek communion with higher powers, it seems difficult not to have a good day. Zen Master Futomaki said, "One does not climb to attain enlightenment, rather one climbs because he [or she] is enlightened." Climber Robert Pirsig adds, "The only Zen you find on the tops of mountains is the Zen you bring up there." I had another such enlightening, Zen-filled day on Grays and Torreys with Bryan Nielson (a high school friend visiting from Bozeman, Montana) two weeks prior to climbing La Plata Peak.

Grays Peak (14,270 ft.) is the 9th highest mountain in Colorado and the county highpoint for Clear Creek County and Summit County. It's also the highest peak in Colorado's Front Range. At age 62, Asa Gray–known as the "Father of American Botany"–made his first visit to Colorado to climb the fourteener that bore his name. Asa, along with his wife and several friends, climbed Grays Peak in 1872. Explorer Charles Parry (1823-1890) named the peak after Asa in 1862.[234] Parry was a British-American botanist and mountaineer, and Colorado's Parry Peak (13,391 ft.) is named after him.

The "walk-up" route on Grays Peak is the Class 2 Stevens Gulch Trail, the same one used by early Colorado tourists who rode horses to the summit. The ride was described in brochures as "an easy day for a lady."[235] Next-door neighbor Torreys Peak (14,267 ft.) is a walk-up from Stevens Gulch too. The standard route follows the Grays Peak Trail to the Grays/Torreys saddle (13,700 feet), and then turns north and ascends Torrey's south slope.[236]

Grays and Torreys are two of the few 14ers I didn't do solo, and while I enjoyed climbing with Bryan, over the years I've come to relish hiking, climbing, hunting, camping, and traveling on my own. I realized early in life that if I was ever going to really see and experience the country and world, I'd most likely have to do it solo. For me, such unaccompanied endeavors are the most satisfying actually. John Muir, being accustomed to roaming alone in the mountains, once wrote: "Oftentimes we are nearest our friends when furthest from them."[237]

Jack Turner, author of *The Abstract Wild*, adds: "Much of what I have learned about myself I learned alone in the wild. The variety of solo adventures is huge, and so are the rewards, but no matter how you go about it, there is always something to be learned... For many, a solo trip is not even an option worth

---

233 *Outside* magazine. "Viesturs's Media Star Rises as Annapurna Looms." *Outside*: 3/22/02.

234 http://www.fourteenerworld.com/

235 Walter R. Borneman. *14,000 FEET: A Celebration of Colorado's highest mountains.* Pueblo, Colorado: Skyline Press, 2005.

236 http://www.summitpost.org/show/mountain_link.pl/mountain_id/229.

237 Kit Stolz. "A Legacy With Legs." *Sierra*: November/December 2001, p. 72.

considering. And it's true that going alone is more dangerous than going with groups. Everything is at stake. But that's the point: You pay attention. You feel more alive."[238]

Adventure traveler (and author) Paul Theroux knows the essence of travel is discomfort and self-discovery. "Real travel is seldom enjoyable at the time," he says, "it's enjoyable in retrospect. Travel is solving problems yourself and overcoming difficulties. It's all about being solitary." Theroux says (in "A traveler, but never a tourist") when you travel with someone else, everything is a plan. You're thinking, "Is my wife comfortable? Happy? Does she want to go shopping? That's a distraction."[239]

Edward Abbey likes to avoid distractions too: "Most of my wandering... I've done alone. I find that in contemplating the natural world my pleasure is greater if there are not too many others contemplating it with me."[240] Unfortunately, a weekend afternoon on a Front Range fourteener is a far, far cry from "getting away from it all," and there are usually droves of other hikers on Grays and Torreys, as described here by Colorado Springs *Gazette* outdoors columnist R. Scott Rappold:

"In 2009, according to the Colorado Fourteeners Initiative, some 750,000 people will visit one of the state's 54 mountains above 14,000 feet. On a Saturday in late June, it seemed a good portion of them were with me on the trail to Grays and Torreys peaks, two hugely popular fourteeners just a few miles from Interstate 70, close enough for Denverites to attain without the discomforts of overnight camping.[241]

"Despite overcast skies, brutal winds, lingering deep snow and clouds obscuring the summits, a veritable conga line of hikers in jeans and cotton stretched down the trail, hundreds of weekend warriors and their dogs transforming an intimate mountain experience into the [Boulder] 16th Street Mall."[242]

Vail Daily contributor Rick Spitzer (in "Easy access to fourteeners") adds, "While climbing Grays Peak a few years ago, I passed 25 people headed down who already made the summit, then I ate lunch with 75 on the summit, and passed another 175 people still ascending the peak on my way back down. This was not a guess—I counted them! It is difficult to think about solitude when expe-

---

238 Jack Turner. "The Solitary Way." *Outside*: April 2003.
239 Paul Theroux. "A traveler, but never a tourist." *USA Today*: 5/19/00, p. 5D.
240 Edward Abbey. *Desert Solitaire*. New York: Simon & Schuster, 1968, p. 200.
241 R. Scott Rappold. "Backdoor trails take you to the top." *The* [Colorado Springs] *Gazette*: 8/12/09.
242 R. Scott Rappold. "Backdoor trails take you to the top." *The* [Colorado Springs] *Gazette*: 8/12/09.

riencing crowds like this. Fortunately, these massive peaks provide many routes for the climber seeking one less traveled."[243]

I wholeheartedly agree with John, Jack, Paul, Ed, Scott, and Rick, that solo adventures are the most enjoyable, but it's also gratifying to share wilderness experiences with people who feel similarly about wildlands, wildlife, and the plight of humanity. The problem is, there don't seem to be many around. Most (but not all) of my friends are not overly concerned about humanity's troubles and shortcomings, and to them, I feel compelled (in the words of Krishnamurti) to say, "It is no measure of health to be well adjusted to a profoundly sick society."

Mountain Gazette contributor (and BHA member) Mark Richards adds, "I'm afraid our big brains might have us heading toward some Kurt Vonnegut future where humans eventually devolve into listless, flatulating seals. Ironically, the very thing that allowed us to survive and prosper as a species—our remarkable adaptability—is the same thing that is doing us in. We have adapted to breathe foul air, drink chemically treated water and eat food that isn't natural or healthy. As the man says, 'So it goes.'"[244]

One of my tests for discovering intelligent and enlightened people is to survey how many books they have in their homes as compared to the size of their so-called "entertainment centers." Generally (but not always), the larger the entertainment center, the fewer the books, the less enlightened the people. Try it for yourself sometime. I think you'll agree it works, unless you're one of those who have already traded reading and learning for sitting zombielike in front of the television.

There is nothing more mindless than regular TV watching. On this point and many others, I agree wholeheartedly with David Petersen, who says: "I haven't watched TV for 35 years, and I can't say I've missed a thing. Supine on my deathbed reflecting back on my small life, I doubt I'll exclaim, 'Damn, if only I'd *watched more TV!*'"[245] By now, I've probably rattled a few cages, but like Ed Abbey said, "If there's anyone still here whom I've failed to piss off, I apologize." On with the show.

Surely there are many good people who don't read much, routinely watch TV, and still manage to lead rewarding lives without ever caring about the fate of humanity or wanting to experience the silence and serenity of wild places and wide-open spaces, but I don't count myself among them. For me, a fulfilling life is one immersed in the outdoors and wilderness, in tranquility and solitude, far away from what we call "civilization," the belch and bile of cities and suburbia.

    I prefer the peaceful silence and shelter of wilderness to the sickening grime

243 Rick Spitzer. "Easy access to fourteeners." *Vail Daily*: 9/12/09.
244 Mark Richards. "The new environmentalism." *Mountain Gazette*: March 2006.
245 David Petersen. *On The Wild Edge*. New York: Henry Holt and Company, 2005, p. 29.

and gridlock of city life. I relish mountain breezes and detest megalopolis mal-
aise, and the only reason for leaving such solitude—to live amongst the urban
masses—is to fight for its preservation. But the battle is getting harder by the minute, mostly due to simple-minded, insidious population growth combined with
gluttonous overconsumption. As famed environmental activist David Brower
said, "Population is pollution spelled inside out."[246]

In his book *Elkheart* (a "love song to elk and elk country"), David Petersen
describes a cartoon David Brower would appreciate: two buck deer are watching a hunter in the distance, and one buck remarks to the other, "Why don't they
have hunting seasons to control their *own* population?"[247] David's friend Edward
Abbey adds, "One must speak plainly on this question: any couple who have
more than two children are not only placing an unfair burden on their neighbors
and on society in general, they are also placing these numerous offspring in a
disadvantageous position in the scramble for survival."[248]

> "It is really not the wilderness that needs
> management (it has been doing quite well, after
> all, for a couple billion years), but people."
> —Roderick Nash[249]

If you're unwilling or unable to sensibly limit yourselves to two, one, or no
children, then at least let your elected legislators and other decision makers know
you want your kids to grow up in a world filled with wild places and wide-open
spaces. Tell them you want to live in a world still populated with spotted owls,
peregrine falcons, California condors, timber wolves, grizzly bears, caribou, lynx,
and all the other species now endangered because of human overpopulation,
overconsumption, out-of-control oil and gas drilling, pervasive shortsightedness,
and extreme stupidity.

Albert Einstein once said it's the responsibility of every human being to aspire
to do something worthwhile, to make this world a better place than the one he
or she found, and in the words of Winston Churchill, "We make a living by what
we get, but we make a life by what we give." Ted Turner is a man who's taken
Einstein's and Churchill's words to heart. Ted gives a lot, he's definitely part of
the solution, and he puts his money where his notoriously big mouth is.

"My grandfather was a farmer," Turner says. "Remember what Scarlet
O'Hara's father said to her: 'The land's the only thing in the world worth work-

246 John McPhee. *Encounters with the Archdruid*. New York: Farrar, Straus and Giroux, 1971, p. 42.
247 David Petersen. *Elkheart*. Boulder, CO: Johnson Books, 1998, p. 131.
248 David Petersen (ed.). *Postcards from Ed*. Minneapolis: Milkweed Editions, 2006, p. 82.
249 Roderick Nash. *Wilderness And The American Mind*. Yale University Press: 1967.

ing for, worth fighting for, worth dying for, because it's the only thing that lasts.'" Turner's actions speak even louder than his words, especially when he uses his considerable wealth to buy land, specifically to preserve species—human and otherwise.[250]

On top of the $500 million he'd already spent as of 2003, Turner spends millions more to restore native wildlife and reverse human-inflicted damage. In "Ted Turner is a worried man," *Fortune* magazine contributor Patricia Sellers explains that Turner's vast land holdings—26 properties in 10 states and two countries—serve to restore the man as well as the wildlands and wildlife he's fighting to save.[251] As of July 2000, Turner had amassed more land in America than any other individual; more than 1.7 million acres. If stitched together, his holdings would be larger than Yosemite, Grand Teton, Rocky Mountain, and Canyonlands National Parks combined.[252]

Despite Turner's extraordinary land purchasing and preservation efforts, as of 2000, the nation's largest private landowner was a paper company, International Paper (IP), which had 7.3 million acres. By 2005, another private company, real estate investment trust Plum Creek Timber, had surpassed IP by amassing 8.2 million acres.[253] Because companies like IP and Plum Creek are purely greed-driven ventures, they're also systematically liquidating their land holdings, selling out to private investors for a quick buck, who, in turn, develop subdivisions and strip malls for their own tidy profit. We can only hope some of them turn out to be more like Ted.

In "Ted Turner builds the ultimate preserve," *USA Today* contributor Patrick O'Driscoll says Turner's ranches are a vast canvas on which he is painting a new vision of land management in the West. Turner says, "We're just getting out of nature's way. That's all we want to do. Get out of the way and let nature go back to the way it was," which means removing much of the old human foot-print. After Turner buys a ranch, his crews remove hundreds of miles of barbed wire fence, decrepit buildings, rusting equipment, non-native weeds and grasses, and land-trampling, stream-degrading cattle.[254]

Then he tries to bring back the natives, from songbirds to bison. O'Driscoll says where developers would eagerly build scores of trophy homes, Turner wants to preserve trophy views and restore endangered species. He'd like to save the planet from overpopulation and urban sprawl and the loss of key pollinators responsible for one-third of the world's food. Because he can't take global matters

---

250 Patricia Sellers. "Ted Turner is a worried man." *Fortune:* 6/2/03, p. 128.

251 Patricia Sellers. "Ted Turner is a worried man." *Fortune:* 6/2/03, p. 128.

252 Patrick O'Driscoll. "Ted Turner builds the ultimate preserve." *USA Today:* 7/17/00, p. 1A.

253 John Ritter. "An intriguing new development in the forest." *USA Today:* 12/23/05, p. 16A.

254 Patrick O'Driscoll. "Ted Turner builds the ultimate preserve." *USA Today:* 7/17/00, p. 2A.

into his own hands, Ted works his vision on his own land, and hopes the example spills over the property and political lines.[255]

But Turner is much more than just a wealthy entrepreneur-conservationist; he's also a hunter who allows paying sportsmen and women to shoot quail, bison, elk, antelope, wild turkey, and other species on his properties. "You need good healthy land for good healthy animals," he says. "They need good water, good cover, and good food. If you're missing any one of those three things, you won't have animals. I maintain my ranches with wildlife being the top priority. I am trying to do the smart thing for the environment instead of the dumb thing. I want others to see what can be done with the land—even if they're not billionaires."[256]

Turner has even found a way for hunting to pay for conservation on his ranches, just like it does for the vast majority of state wildlife agencies. At his Vermejo Park Ranch, which straddles the New Mexico-Colorado border in the Sangre de Cristo Range, Turner allows hunters to kill 200 trophy elk each year—about 2 percent of his 10,000-elk herd. Each hunter pays $10,000, which generated two million dollars in revenues annually. "Now, that's a pretty acceptable figure," says Turner, who uses the income to keep his 600,000-acre property in a wild and natural state, with few fences and preference given to indigenous plants and animals.[257]

Going forward, whatever he does or doesn't do to sway the fate of humanity, Ted Turner will keep on moving, says Patricia Sellers. "You should set goals beyond your reach so you always have something to live for." That's what his father told him, she explains. "You never have enough time in life," Turner adds. "I'm constantly battling to stay ahead. I say to myself, 'All you have to do is put one foot in front of the other and just keep walking.'"[258] Just like climbing a mountain.

Like Turner, I've never been one to follow in anyone's footsteps for very long, and at the tender age of 24, I realized if I was ever going to see and experience the world, truly broaden my horizons beyond the standard high school-college-marriage-house-kids-grandkids-nursing home life path, I'd likely have to do it on my own; blaze my own trail through life. Always waiting for someone else (be they friends, family, girlfriends, or wives) to do things with you is a sure recipe for seeing, experiencing, and accomplishing little or nothing.

Henry David Thoreau once said, "Men frequently say to me, 'I should think you would feel lonesome down there, and want to be nearer folks.' I am tempted

---

255 Patrick O'Driscoll. "Ted Turner builds the ultimate preserve." *USA Today*: 7/17/00, p. 2A.
256 Robert M. Poole. "Hunters: For Love of the Land." *National Geographic*: November 2007.
257 Robert M. Poole. "Hunters: For Love of the Land." *National Geographic*: November 2007.
258 Patricia Sellers. "Ted Turner is a worried man." *Fortune*: 5/26/03, p. 132.

to reply to such,—This whole earth which we inhabit is but a point in space. How far apart, think you, dwell the two most distant inhabitants of yonder star, the breadth of whose disk cannot be appreciated by our instruments? Why should I feel lonely? Is not our planet in the Milky Way?"[259] Rachel Carson adds, "Those who dwell among the beauties and mysteries of life are never alone or weary."

Dorothy Molter—the famous Boundary Waters Canoe Area Wilderness "Root Beer Lady" who lived on remote Knife Lake from 1930 (when she was 23-years-old) until her death in 1986—thrived on wild solitude too: "I live alone, but I'm not lonely."[260] To those like Ted, Henry, Rachel, and Dorothy, life means adventure and exploration. It's about getting up and moving on; pushing the edge of the envelope; taking risks. To live is to risk, and the person who takes no risks does not really live: Even a dead fish can go with the flow.

"Why even bother to swim upstream? Why not just go with the easy flow?" asks (and answers) David Petersen. "Good questions—to which gonzo journalist and political pundit Hunter S. Thompson supplies answer enough... 'This is the real point: that we are not really freaks at all... [Rather,] the twisted realities of the world we are trying to live in have somehow combined to make us feel like freaks. We argue, we protest, we petition—but nothing changes.'"[261]

Canyon country vagabond Everett Ruess was a go-against-the-flow type too. "I have been thinking more and more that I shall always be a lone wanderer of the wilderness. God, how the trail lures me. You cannot comprehend its resistless fascination for me. After all the lone trail is the best... I'll never stop wandering. And when the time comes to die, I'll find the wildest, loneliest, most desolate spot there is."[262] Everett's brethren, Teddy Roosevelt, adds, "The farther one gets into the wilderness, the greater is the attraction of its lonely freedom."

Polar explorer Will Steger said, "An adventurous nature demands: Accept the dare. Take the next step. Onward, upward. There is a spark in the adventurous spirit that won't be extinguished, a thirst never slaked. Death has ended the career of many adventurers, yet one is sure that most of them died without regret and would otherwise have continued on to new adventures the next day."[263] Like Everett, Teddy, and Will, I refuse to live life with more than a few regrets, and I will not ignore the fact that each of us must take responsibility for the future of our planet if it's going to be left even marginally intact for future generations.

Humans are the only species able to go everywhere in the world, which also means we have the capacity to do good or spread ill everywhere. We are by

259 Amy Kelley (ed.). *Inspired by Nature.* Helena, MT: Falcon Publishing, Inc., 2000, p. 100.

260 Dennis Anderson. "BWCAW sled dogs offer cold comforts." [Minneapolis-St. Paul] *Star Tribune*: 3/3/10.

261 David Petersen. *On The Wild Edge.* New York: Henry Holt and Company, 2005, p. 243.

262 John Nichols. *Everett Ruess.* Layton, UT: Gibbs M. Smith, Inc., 1983, p. 78.

263 Will Steger. "The Spirit of Adventure." *The Greatest Adventures of All Time.* Des Moines, IA: LIFE Books, 2000, p. 9.

far the most adaptable mammals on the planet. As *Anchorage Daily News* outdoors columnist Craig Medred says, "We survive in places where rats can't even make it." We function "in climates that cause cockroaches to expire," and we can populate and decimate the entire planet if we choose, if we believe that our superior adaptability and technology trumps the right of every other living thing on earth to exist.[264]

Just because we currently have the ability to fling space stations into the heavens, split atoms, manufacture toxins, and cause mass destruction everywhere, does that give us the right? Ingenuity is one thing, wisdom another. Once we bulldoze, develop, drill, log, mine, and overrun most of the world's remaining wildlands, we will have done the unthinkable. We will have destroyed our evolutionary homes, mangled creation—the heart and soul of humanity—and we will pay the price. Humankind has never been so dangerous or vulnerable. "For 200 years we've been conquering Nature. Now we're beating it to death," says Tom McMillan, former Canadian Minister of the Environment.[265]

Nature Conservancy contributor Katherine E. Reed counters with: "I have the utmost confidence in humanity's ability to solve a problem once we decide it's truly a problem."[266] Maybe the hardest case to make for taking action and solving our nation's and world's many woes is that it's, quite simply, the right thing to do, and if we don't do it, we'll face problems of epic proportions down the line, which will most likely be our collective undoing. Yet, in the end, it may be the only one worth making. If we don't protect and respect nature and all the other species on this ark called Earth, we truly do not respect ourselves.

On May 2, 1899, when Theodore Roosevelt had been New York governor for only four months, he signed into law an educational initiative requiring schools to teach natural history and geography. Young citizens, he believed, needed to understand the evolutionary process and learn why dumping sewage and refuse into the Great Lakes and Hudson River was unacceptable. In a sense, promoting Earth Day seventy years ahead of its time, explains David Brinkley in *The Wilderness Warrior*, Roosevelt believed that humans couldn't afford to recklessly poison their own environment without incurring a heavy toll in ill health, environmental ugliness, and corrosion of the spirit.[267]

As Rachel Carson said, the question is whether any civilization can wage relentless war on life without destroying itself and without losing the right to be

---

264 Craig Medred. "Treadwell lived life in vastly different worlds." *Anchorage Daily News*: 10/26/03.

265 George Wuerthner (ed.). *Thrillcraft*. White River Junction, Vermont: Chelsea Green Publishing Company, 2007, p. 180.

266 Katherine E. Reed. "Chief Environmental Officer–the *Other* CEO." *Nature Conservancy*: Winter 2002, p. 50.

267 Douglas Brinkley. *The Wilderness Warrior: Theodore Roosevelt And The Crusade For America*. New York: HarperCollins Publishers, 2009, p.352

called civilized. In *Visions of Wild America*, Kim Heacox notes that along with Thoreau, Muir, Roosevelt, and Leopold, "Carson" is one of the few surnames in American conservation history that stands by itself. Her work is a testament to the fact that illumination comes from those who differ, rather than agree, with the prevailing order. Rachel Carson's indictment of pesticides, herbicides, and the powerful industries that produce them in her book *Silent Spring* shook up the prevailing order and caused an uproar.[268]

The truth often does when it threatens the political, corporate, or societal status quo. According to Heacox, the storm Rachel created was immediate and fierce, including parodies and personal attacks on that "hysterical woman" and "nun of nature." *Time* magazine labeled her book an "emotional and inaccurate outburst." After all, the man who'd discovered the insecticidal properties of DDT won the Nobel Prize. Even the American Medical Association was upset, which is ironic when you consider that a doctor's oath is, "First, do no harm."[269]

Even though her personality was low-key (explains *Desert Morning News* contributor Dennis Lythgoe), Carson became the spokeswoman for a network of conservationists, scientists and others who wanted to stop what they saw as the "human assault on nature."[270] Carson's attackers knew she was questioning not just the use of poisons, but the basic irresponsibility of our industrial society toward the natural world: the belief that damage to nature is an inevitable and acceptable cost of "progress." That was her crime, her heresy, but it was one that changed the world.[271]

The chemicals and pesticides industries, the media, government scientists and politicians, all of them attacked Carson, portraying her as a hysterical woman; her facts were called inaccurate, her competency questioned, her good name bashed, but her book became a best-seller.[272] Rachel's lyric calm (explains *Onearth* contributor David Gessner) in the pages of *Silent Spring*, and her professional composure while testifying before Congress, belied the efforts of chemical companies to smear her as a hysteric and, if you can believe it, a Communist sympathizer.[273] Anais Nin (a U.S. author and diarist) said, "Life shrinks or expands in proportion to one's courage," and Rachel was as courageous as they come.

In her *National Geographic* article, "Environmental Movement at 40," Hillary Mayell explains that the "silent spring" was Rachel's metaphor for human-caused

---

268 Kim Heacox (ed.). *Visions of a Wild America*. Washington, D.C.: National Geographic Society, 1996, p. 115.

269 Kim Heacox (ed.). *Visions of a Wild America*. Washington, D.C.: National Geographic Society, 1996, p. 115.

270 Dennis Lythgoe. "Rachel Carson in spotlight." *Desert Morning News*: 2/11/07.

271 Kim Heacox (ed.). *Visions of a Wild America*. Washington, D.C.: National Geographic Society, 1996, p. 115.

272 Hillary Mayell. "Environmental Movement at 40: Is Earth Healthier?" *National Geographic News*: 4/20/02.

273 David Gessner. "A New Climate for Literature." *Onearth*: Fall 2009, p. 24.

environmental destruction: A small American town, the farms prosperous, the people healthy and happy, the forests teeming with wildlife, all silenced, sickened by a mysterious illness brought on by the people themselves. The "strange blight" Carson described was caused by the unexamined and virtually unregulated practice of dumping, spraying, dusting, and otherwise distributing deadly chemicals into the environment.[274]

"So you're the little lady who started this whole thing," Abraham Lincoln said during the Civil War to Harriet Beecher Stowe, author of *Uncle Tom's Cabin*. A century later, Rachel Carson appeared before Senator Abraham Ribicoff's committee amid another war–the one against environmental toxins. "Miss Carson," Senator Ribicoff said, "you are the lady who started all this."[275] What she started accomplished one of the most difficult tasks imaginable; it changed minds. It transformed the way we think about ourselves and the world around us, and it saved a symbol.

"All truth passes through three stages. First, it
is ridiculed. Second, it is violently opposed.
Third, it is accepted as being self-evident."
–Arthur Schopenhauer

The embodiment of American strength and freedom, the bald eagle is our nation's symbol. As told by *National Geographic* contributor John Eliot (in "bald eagles come back from the brink"), eagles ruled the skies on seven-foot wingspans when seventeenth-century Europeans arrived in North America. Across the continent half a million bald eagles may have soared, but settlers blamed them for killing livestock, so the shooting began and the proud birds' numbers plunged. Alaska's territorial legislature even enacted a bounty on eagles in 1917. Finally, 23 years later (in 1940) the Bald Eagle Protection Act was passed by Congress.[276]

The act prohibited shooting or otherwise harming eagles in the U.S., but didn't address the pesticides that, within a decade, began destroying eagles' eggs by weakening their shells. By the 1960s, only about 400 breeding pairs of bald eagles remained in the Lower 48 states. "The trend... may well make it necessary for us to find a new national emblem," Carson warned in *Silent Spring*.[277] Widespread use of the pesticide DDT was decimating eagle and other bird populations nationwide at the time, not to mention its cancerous impact on humanity. Then a 1972 ban on DDT, along with protections offered by the

274 Hillary Mayell. "Environment Movement at 40: Is Earth Healthier?" *National Geographic News*: 4/20/02.
275 Kim Heacox (ed.). *Visions of a Wild America*. Washington, D.C.: National Geographic Society, 1996, p. 123.
276 John L. Eliot. "bald eagles come back from the brink." *Selections from National Geographic*: 2004, p. 42.
277 John L. Eliot. "bald eagles come back from the brink." *Selections from National Geographic*: 2004, p. 42.

Endangered Species Act (passed in 1973), set the stage for a comeback.

Today, there are an estimated 7,000 eagle pairs in the Lower 48 states. Rachel would be proud, but also dismayed by how much work there's still to do. In 1992, the same year a panel of distinguished Americans voted *Silent Spring* the most influential book of the last half century, 2.2 billion pounds of pesticides were used in the United States—eight pounds for every man, woman, and child.[278] No wonder so many (maybe most) of us will be diagnosed with some form of cancer during our lives.

Ironically and tragically, Rachel Carson died of cancer in 1964, an affliction most likely caused or exasperated by carcinogenic pesticides like DDT. Rachel was posthumously awarded the Presidential Medal of Freedom for her groundbreaking work, but since her death, nationwide cancer rates have skyrocketed. Between 1974 and 1994, cancer deaths in the U.S. increased 49 percent, from 358,881 to 534,300.[279]

MD Tom Gritzka lays out the deadly and daunting statistics: "The incidence of breast cancer in the U.S. has gone from 1/200 when I was in medical school to 1/6 today—from 0.5% to 16%. Prostate cancer has doubled since 1950. Testicular cancer has increased 100% since 1950; in males in the age range from 25-45, it has increased 300%. Male breast cancer, once a rarity, is now common... [George] Bush and his ilk worship Mammon, not Jesus, as they claim. They are killing us for corporate profit."[280]

To paraphrase *High Country News* contributor Auden Schendler (praising climatologist James Hansen, in "When a scientist becomes an activist"), Rachel Carson could have been content with just being a scientist. She could have done her work and felt—with justification—that she had contributed greatly to the world. This would have freed her from the personal attacks, the stress, and the burden of being a living Cassandra, determined to try to change a world that stubbornly refuses to listen. But Carson elected not to do that. She chose, instead, to be a great woman.[281] We need many more women, and men, like her. Thanks, Rachel, and James.

OCTOBER

Rachel Carson, James Hansen, Dr. Gritzka, and Edward Abbey weren't afraid to speak their minds or stray from the herd. In his book *Desert Solitaire*, Ed Abbey writes about the redrock canyon country of southern Utah and breaking away

278 Kim Heacox (ed.). *Visions of a Wild America.* Washington, D.C.: National Geographic Society, 1996, p. 123.
279 American Cancer Society. "Cancer Deaths Rise." *Business Week:* 6/1/98, p. 146.
280 Tom Gritzka, M.D. "Write to REP..." *The Green Elephant:* Spring 2005, p. 12.
281 Auden Schendler. "When a scientist becomes an activist." *High Country News:* 3/11/10.

from the herd: "Man is a gregarious creature, we are told, a social being. Does that mean he is also a herd animal? I don't believe it, despite the character of modern life. The herd is for ungulates, not for men and women and their children."[282] Like Ed, I have endeavored to do more than follow the herd during my life.

And if you live in Colorado, that frequently leads to climbing mountains and exploring the strange and breathtaking redrock canyon country of nearby southern Utah. After hiking-climbing La Plata Peak on September 25, I didn't summit another 14er until November, partly due to taking time off during October for an annual but preferably more frequent pilgrimage to the redrock canyons of southern Utah, the region Edward Abbey spent most of his life exploring, writing about, and tirelessly defending.

Abbey described the surreal and vast canyonlands of the Colorado Plateau as, "The least inhabited, least inhibited, least developed, least improved, least civilized... most arid, most hostile, most lonesome, most grim bleak barren desolate and savage quarter of the state of Utah–the best part by far."[283] Ed's friend and fellow canyon crawler David Petersen says to "experience this region, for the hidden treasure it is, you must abandon your wheels and explore its hundred twisting miles of trail by boot or saddle. The ringing *quiet* you'll meet down there alone justifies the effort."[284]

One of my favorite canyon country destinations, the Maze District of Canyonlands National Park, is also one of the most remote and awesomely beautiful places left in the Lower 48–the type of country the term *terra incognita* still applies to. The canyonlands are part of the much-larger 108-million-acre Colorado Plateau, one of the world's most distinctive landscapes. The Colorado Plateau covers 13,000 square miles in northern Arizona, western Colorado, northwestern New Mexico, and eastern Utah.[285] These Four Corners states collectively contain 46 national parks and monuments, 23 national forests, and more than 56 million acres of Bureau of Land Management lands.[286]

Although not known for mountains, the Colorado Plateau has its fair share. The La Sal Mountains are located on the eastern border of Utah about 15 miles southeast of Moab. Rising over 7,000 feet above the surrounding redrock canyons and mesas, the La Sals harbor the highest peak on the Colorado Plateau, 12,721-foot Mount Peale.[287] Although this range occupies a relatively small

282 Edward Abbey. *Desert Solitaire*. New York: Simon & Schuster, 1968, p. 58.

283 Tyler Williams. "Henry Mountains: Skiing The Desert." *Inside/Outside Southwest*: January/February 2008.

284 David Petersen. *Cedar Mesa: A Place Where Spirits Dwell*. Tucson: The University of Arizona Press, 2002, p. 12.

285 Douglas Brinkley. *The Wilderness Warrior: Theodore Roosevelt And The Crusade For America*. New York: HarperCollins Publishers, 2009, p. 528.

286 National Geographic Adventure (NGA). "21st-Century Adventurers." *NGA*: March 2007, p. 72.

287 I climbed Mount Peale during June 2007.

area—15 miles north to south and six miles across—from Mount Peale's summit, to the east, you gaze upon the high peaks of the San Juan Mountains, in Colorado, and to the west Utah's canyon country with the Henry Mountains in the far distance.[288]

Another peak I'd like to climb in the La Sals, Mount Tukuhnikivats (12,489 ft.), is one Edward Abbey climbed while working as a park ranger in nearby Arches National Monument (now a national park). "The land is lovely now, more beautiful every day," Ed wrote. "The golden light of autumn is beginning to appear, not in the sky but in the flowers: matchweed, rabbitbrush, princess plume, beeweed, mule-ear sunflowers, all are flaming forth in yellow. A cold October wind blows down from the mountains, tho' it's still September here below. I must climb Tukuhnikivats once more before I leave."[289]

The Colorado Plateau started to take form tens of millions of years ago when a low-lying area of boggy inland seas was lifted upward by tectonic forces through a vertical mile, exposing the soft sediments below to the canyon-cutting power of what would become the Colorado River. At various times, magma thrust up through the sedimentary rocks to form high, islandlike mountain ranges. Today, this dramatic geologic display has been recognized by the designation of over 30 national parks and monuments, the greatest density of such protected areas on earth, according to the Grand Canyon Trust.[290]

In 1961, Interior Secretary Stewart Udall visited remote reaches of the Colorado Plateau in southern Utah and began lobbying Congress for the creation of a national park in the midst of this redrock wilderness. Three years later, Canyonlands National Park was born. Stretching into the state's southeast corner, the park consists of deep canyons, grand gorges, sturdy buttes, and ornate spires shooting high into the bright blue sky, and at 527 square miles, it's the largest national park in Utah.[291]

Canyonlands is a biblical wilderness of redrock peaks, canyons, cliffs, towers and arches, the likes of which are found nowhere else in the world. A place of wild wonder and worthwhile worship if there ever was one. Step into Canyonlands National Park, and you enter perhaps the most rugged backcountry park in the Lower 48, says *National Parks Traveler* contributor Kurt Repanshek. Though it covers just 337,598 acres, or less than one-seventh the size of 2.2-million-acre Yellowstone National Park, Canyonlands feels much larger.[292]

---

288 http://cpluhna.nau.edu/Places/lasalmts.htm.
289 David Petersen (ed.). *Postcards from Ed*. Minneapolis: Milkweed Editions, 2006, p. 12.
290 Grand Canyon Trust. "Projects of the Colorado Plateau," 5/30/08.
291 http://away.com/ideas/southwest/utah_canyonlands.html.
292 Kurt Repanshek. "Will The Long-Desired "Completion" Of Canyonlands National Park Ever Arrive?" *National Parks Traveler*: 2/21/10.

The Green and Colorado rivers intersect at the heart of Canyonlands, creating what's known as the River District, the first of the park's four sub-regions. Island in the Sky, essentially a massive plateau rising between the river valleys, lies to the north. The remote Maze District is to the west and the Needles District, named after its colorful rock spires, to the east.[293]

Hands down, this is the heart and soul of America's wild public lands heritage and my favorite place in the world. "Crazy country," Abbey wrote in his '70s eco-cult classic *The Monkey Wrench Gang*. A desert territory as much vertical as it is flat, so gashed and shoved about by time, tectonics, and weather that it's daunting and inaccessible even to someone on foot. *Onearth* contributor Alex Shoumatoff says, "It is a Zen landscape, sculpted by chance and the laws of cause and effect, by processes that have been going on for millions of years in which we are nothing."[294]

"Whoever named it knew it," remarked veteran canyon country guide Kent Frost in his 1971 memoir *My Canyonlands*.[295] Kent was born and raised in southern Utah and spent most of his life exploring its wonderland of canyons and otherworldly terrain. About viewing the Maze District from atop what is now Canyonlands National Park's Flint Trail, he wrote: "We gazed for a long time at the vast land of rainbow rock confusion below us. We saw ragged escarpments in the distance. Far across the Colorado, to our east, the Needles country was backed by shadowy mountains. The wind-and-rain-carved pinnacles were calling us from down below."[296]

They called me too, gazing down into the Maze from that same spot on October 4, 1999. Below was a territory of multihued cliffs, plateaus, mesas, buttes, pinnacles, and canyons spread out for as far as the eye could see. In the words of William B. Smart, "It's big country out there. With a topo map and a beat-up jeep—or better yet, a backpack—there's still solitude to be found, adventure with the smell of danger, and the wonder of a raw, unfinished, and awesomely beautiful world," but to enter it, I had to negotiate the Flint Trail.

The Flint is a merciless "road" that was also an escape route for the heroes in Ed Abbey's book *The Monkey Wrench Gang*. Unless you're highly skilled at technical four-wheeling, as the National Park Service personnel at the Hans Flat Ranger Station warned me, the top of the Flint Trail is a good place to park. I considered (but ignored) their well-meant advice, and will never again enter the Maze unless on foot, horseback, or river raft. But I'm glad I went, because the Maze is

---

293 http://away.com/ideas/southwest/utah_canyonlands.html.
294 Alex Shoumatoff. "Who Owns This River?" *Onearth*: Spring 2005, p. 27.
295 Ray Boren. "Under the ledge: Park Service strives to keep the 'wild' in the tangled wilderness of Canyonlands' remote Maze District." *The Desert News*: 8/3/00.
296 Kent Frost. *My Canyonlands*. New York: Abelard-Schuman, 1971, p. 124.

one of the wildest and most breathtaking places in, dare I say, the entire world.

The folks at Get Outdoors (www.getoutdoors.com) explain the challenges one faces upon entering this redrock wilderness: "Due to the district's remoteness and the difficulty of its roads and trails, the Maze is the least accessible district of Canyonlands, requiring more time as well as a greater degree of self-sufficiency to visit. Trips usually last a minimum of three days, and the area can easily absorb a week-long trip. Maze visitors should be skilled in the use of topographic maps and, if traveling by vehicle, in technical four-wheel driving."[297]

During 1999, over 446,000 canyon country tourists and adventurers visited Canyonlands National Park. More than half marveled at the high vistas from the Island in the Sky District, easily accessible using paved and four-wheel-drive roads from Moab. The Needles District, between Moab and Monticello to the southeast, had the second-highest number of visits, but only about 2 percent of the park's visitors—10,613 during all of 1999—made their way into the Maze; Canyonlands at its wildest.[298] I was one of them.

> "They did that? For what? They went
> there? Why? Because it was there."
> —Robert Sullivan[299]

Former Southern Utah Wilderness Alliance (SUWA) board member Ray Wheeler tells us why the remaining redrock wilderness of southern Utah is so valuable. Why it must be saved for current and future generations: "Yet the remoteness, the size, the harsh terrain, the heat, the aridity... the wind, the silence when there is not wind, the overpowering solitude... these are precisely the commodities which make it valuable. It is a fierce and dangerous place, and it is wilderness right down to its burning core."

It's also at grave risk, because the Bush administration opened up most of our remaining unprotected roadless public lands in southern Utah and everywhere else to ATVs, miners, and oil and gas drillers. For example, during the 2008 Utah BLM resource management and travel plan updates, the Bush administration surrendered 84 percent of our public BLM lands to the oil and gas industry, jeeps, ATVs, and dirt bikes; for motorized wreckreation playgrounds and miniscule amounts of oil and gas.[300]

297 http://www.getoutdoors.com/getoutdoors/destination_regions/5519.html.

298 Ray Boren. "Under the ledge: Park Service strives to keep the 'wild' in the tangled wilderness of Canyonlands' remote Maze District." *The Desert News*: 8/3/00.

299 Robert Sullivan and Robert Andreas (eds.). "Adventure: An Attempt at a Definition." *Life: The Greatest Adventures of All Time*. Des Moines, IA: Time, Inc., 2000, p. 11.

300 Liz Thomas. "Bush BLM Leaves a Destructive Legacy in Utah's Canyon Country." *Redrock Wilderness*: Autumn/ Winter 2008.

When Edward Abbey was a ranger at Arches National Monument in the 1950s (explains *Thrillcraft* contributor Richard Mahler), most Americans were probably repelled by the windswept wilderness of eroded, sun-baked rock that dominates the area. Abbey's biggest fear, according to fellow writer Timothy Egan, was that someday, visitors would run their motorized vehicles through such landscapes, which Abbey considered sacred, in part because of their silence. Some places should remain off-limits forever to motorized vehicles, he believed. "After all," Ed reasoned, "you can't drive a car into church."[301]

And the notion that oil and gas drilling amidst roadless public lands is needed to bring new energy-rich areas into production is a farce. In Utah, for example, 95 percent of oil and gas production comes from seven key regions. All seven of these energy-producing regions are open to oil and gas development. Given that the areas proposed for wilderness designation in Utah contain less than 5 percent of the state's energy resources, drilling in them would amount to no more than a drop in the bucket.

U.S. Geological Survey (USGS) data indicates that the total amount of undiscovered oil in Utah is only enough to supply the U.S. for about three weeks. Even in highly productive parts of the state, 11 percent of wells produce no oil or gas. Outside of these proven areas, 63 percent of wells produce no oil or gas.[302] Against such unfavorable odds, the Bush administration wagered some of the most spectacular wildlands left in the world. The oil to be recovered in Utah would amount to what Americans use in three weeks. Three weeks!

During the Bush years, oil and gas companies, mining conglomerates, and logging syndicates were sweeping across the West like General Sherman's Civil War march to the sea, scraping and cutting, polluting and denuding, all in the name of political payoffs and greed. *Christian Science Monitor* contributor Brad Knickerbocker said ("In Utah, a public-land fight on an epic scale"), the conflict boiled down to world-class scenery and wildlife habitat in a still spacious and wild redrock wilderness in conflict with, at best, very insignificant oil and gas deposits.[303]

"At least the war on the environment is going well."
—Bumper sticker in Moab, Utah[304]

---

301 George Wuerthner (ed.). *Thrillcraft*. White River Junction, Vermont: Chelsea Green Publishing Company, 2007, p. 40.
302 Southern Utah Wilderness Alliance (SUWA). "Drilling in Wilderness: A High Stakes Gamble." *Redrock Wilderness*: Spring 2004, p. 8.
303 Brad Knickerbocker. "In Utah, a public-land fight on an epic scale." *The Christian Science Monitor*: 8/7/03.
304 Southern Utah Wilderness Alliance (SUWA). *Redrock Wilderness*: Autumn/Winter 2007, p. 17.

Oil industry officials claim modern drilling techniques greatly reduce environmental impacts, but when looking out over a once-wild landscape scarred with a maze of motor vehicle tracks, oil pumps, waste pits and pipelines—public lands turned into lifeless industrial zones—Liz Thomas of SUWA (www.suwa.org) asks, "Is this the highest, best value for this land?"[305] Southern Utah's redrock canyon country represents some of the last and largest-remaining desert blank spots on the map—places where towering buttes, sweeping plateaus, and intimate canyons are enveloped by a rare and breathtaking silence.[306] Can't we save what little is left?

A poll conducted by Wirthlin Worldwide, the conservative Republican pollster who did Ronald Reagan's numbers, found that 74 percent of Utahns believe what remains of Utah's wilderness ought to be protected. SUWA's Mike Matz (now with the Campaign for America's Wilderness) said, "That's as close to a mandate as it gets, particularly when one scrutinizes the cross-tabs and finds that the majority of people who responded to the poll have lived in Utah 20 years or more and identify themselves as conservatives. These are *real* Utahns."[307]

In another poll by Dan Jones & Associates (a Salt Lake City polling firm), a solid majority of Utahns said "protecting Utah's wilderness lands should be a high priority for Governor [Jon] Huntsman," with 65 percent giving it a priority rating of 4 or 5 on a 1 to 5 scale and 87 percent a rating of 3 or higher. By that same percentage, Utahns agreed that "there are some public lands where motorized access should not be permitted."[308]

In a document titled *Wilderness at the Edge: A Citizen Proposal to Protect Utah's Canyons and Deserts* (published by the Utah Wilderness Coalition in 1990), Wallace Stegner wrote: "The Utah deserts and plateaus and canyons are not a country of big returns, but a country of spiritual healing, incomparable for contemplation, meditation, solitude, quiet, awe, peace of mind and body... Factories, power plants, resorts, we can make anywhere. Wilderness, once we have given it up, is beyond our reconstruction."[309]

In 2006, at age 86, Stewart Udall remembered as if it were yesterday the moment he first cast his eyes on the rugged country that is now Canyonlands National Park, explains *Salt Lake Tribune* contributor Lisa Church. As President Kennedy's interior secretary, he was flying over southeastern Utah with Floyd Dominy, commissioner of the Bureau of Reclamation, who boasted of plans to

305 Brad Knickerbocker. "In Utah, a public-land fight on an epic scale." *The Christian Science Monitor*: 8/7/03.

306 Orion. "Southern Utah Wilderness Alliance." *Orion Grassroots Network* (www.oriononline.org): 1/24/02.

307 Mike Matz. "The Bigger Wilderness Picture." *Southern Utah Wilderness Alliance*: Fall/Winter 1998, p. 3.

308 Southern Utah Wilderness Alliance (SUWA). "The Good, the Bad, and the Ugly: Utah's Cowboy Caucus Rides On." *Redrock Wilderness*: Summer 2007.

309 Orion. "Southern Utah Wilderness Alliance." *Orion Grassroots Network* (www.oriononline.org): 1/24/02.

build a large dam at the confluence of the Colorado and Green rivers. Within months of that first flight, Udall had organized a much-publicized excursion to see the area from the ground.[310]

Accompanied by then Senator Frank Moss (D-Utah), Arches National Monument Superintendent Bates Wilson, and a throng of reporters and photographers from the nation's largest newspapers and magazines, Udall's trip cast a beacon on southeastern Utah that captured the public imagination and ultimately persuaded Congress to establish what would be Utah's third of five national parks.[311]

As told by *High Country News* contributor Gary Nabhan, "Just a few months after his confirmation in 1961, Udall was disappearing deep into Utah's red-rock country to shape the future Canyonlands National Park with legendary Bates Wilson... at his side. Knowing their jeeps would take them far from civilization for many days, Udall arranged for government memos, the *Washington Post*, *New York Times*, scotch, and ice cream to be air-dropped to his crew every few days. No Facebook or Twitter back then."[312]

Udall counts Canyonlands among his grandest accomplishments, and during his 1961-1968 tenure as interior secretary, he helped protect over 4 million acres of public lands and more than 60 additions were made to the National Park System, including Canyonlands National Park, North Cascades National Park in Washington, Redwood National Park in California, and the Appalachian National Scenic Trail stretching from Georgia to Maine.[313] Udall also wrote a book, *The Quiet Crisis*, a brief environmental history of the United States. We are fortunate to have had such a person among us at a pivotal moment in our nation's history.[314] Thanks, Stewart!

Those were heady times, and we look back to them for guidance and inspiration as we follow in Stewart's footsteps and build on his legacy. Stewart said, "That was a wonderful time, and it carried through into the Nixon administration, the Ford administration, into the Carter administration. I don't remember a big fight between Republicans and Democrats... There was a consensus that the country needed more conservation projects of the kind we were proposing." A consensus that's sadly lacking in the current era.[315]

---

310 Lisa Church. "Udall returns, savors his legacy." *The Salt Lake Tribune*: 7/27/06.

311 Lisa Church. "Udall returns, savors his legacy." *The Salt Lake Tribune*: 7/27/06.

312 Gary Nabhan. "Gary Nabhan remembers Stewart Udall." *High Country News*: 3/29/10.

313 Barry Massey. "Former Interior Secretary Stewart Udall, environmental crusader, dies at age 90." *Associated Press*: 3/21/10.

314 John Worlock. "Life and Accomplishments of Stewart Udall." *Save Our Canyons*: June 2010, p. 6.

315 John Worlock. "Life and Accomplishments of Stewart Udall." *Save Our Canyons*: June 2010, p. 6.

NOVEMBER

After exploring Utah's redrock canyonlands during October, I returned to Colorado. Back to climbing 14ers, including the third-highest peak in the Lower 48 and Colorado's second highest: Mount Massive. Massive rises from the center of the state's tallest mountain range along the Continental Divide. Combined with several other nearby peaks (Mount Elbert, Mount Harvard, and La Plata Peak), Massive puts up a lofty profile that includes four of Colorado's five highest mountains.[316] Mount Massive was "named for its size" in 1873 by Henry Gannett, a member of the Hayden Survey.[317]

Gannett went on to become the USGS's chief geographer from 1882 to 1896, being named to the position by John Wesley Powell. In 1906, Wyoming's Gannett Peak (13,804 ft.), the highest in the state, was named after him.[318] I was privileged to climb Gannett Peak (in July 2007) with a group from the Colorado Mountain Club during a five-day, 40-plus-mile backpack trip into Wyoming's remote and wild Wind River Mountains-Bridger Wilderness. When you tread upon the rock (for you budding geologists) in Wyoming's 428,087-acre Bridger Wilderness Area, you're touching something ancient: 3.6-billion-year-old igneous rock, the oldest in the United States.

Not to be outdone, according to Summitpost.org, Mount Massive has more area above 14,000 feet than any other mountain in the Lower 48 states, narrowly edging out Mount Rainier in this category. Massive, along with Mount Elbert, forms much of the western skyline of Leadville, located 11 miles east and slightly to the north. The standard route up Massive is a 13.6-mile round-trip through 4,370 feet of elevation gain on a Class 2 trail. The mountain and 30,540 acres of surrounding public lands were designated the Mount Massive Wilderness by Congress in 1980.[319]

As described by the Colorado Mountain Club, "From the top of 14,421-foot Mt. Massive, the landscape turns into a polychrome sea of peaks extending in all directions. With all the red, purple, and brownish summits stretching underfoot—the peaks of the Hunter-Fryingpan Wilderness lie due west and the San Isabel National Forest is just south—a lucky hiker just might feel on top of the world. Indeed, it's tough to get much higher. There are just two superior summits in the Lower 48—Mt. Elbert, rising 12 feet taller just to the south, and California's Mt. Whitney, which bests Massive by 74 feet."[320]

The standard Mount Massive approach follows part of the Colorado Trail, which

316 John Fielder and Mark Pearson. *Colorado's Wilderness Areas.* Englewood, CO: Westcliffe Publishers, 1994, p. 201.
317 http://www.fourteenerworld.com/.
318 Roger Rowlett. "Henry Gannett–Father of United States Highpointing." *Apex to Zenith*: 4th Quarter 2007, p. 10.
319 http://www.summitpost.org/show/mountain_link.pl/mountain_id/201.
320 Colorado Mountain Club (CMC). "Mount Massive Wilderness." *Trail & Timberline*: Spring 2010, p. 29.

extends from Denver to Durango crossing seven national forests, six wilderness areas, five river systems, and eight mountain ranges along the way. This nearly 500-mile-long footpath was built by thousands of volunteers for use by those interested in the muscle-powered, quiet-use of some of Colorado's wildest backcountry.

Backpacker magazine says, "The Colorado Trail, running 480 miles from Chatfield State Park (20 miles south of Denver) to Junction Creek trailhead (3.5 miles northwest of Durango), is a life list romp over jagged peaks, saw tooth ridges, and rolling, columbine-filled meadows."[321] "Everest? Pshaw," notes *Backpacker*: you'll gain 77,690 feet along the Colorado Trail. And after crossing over the Continental Divide at 10,800 feet, just above Georgia Pass, you start a high-alpine tour de force—including 38 consecutive miles above tree line—that lasts until your final decent some 300 miles later.[322]

Colorado Mountain Club member Susan Paul adds: "The Colorado Trail is not for the weak of heart or the slight of foot. Rather, it beckons most to those experienced hikers who've grown weary of the day hikes and want more. More mountains, more trail, more vistas and vales. More of the wild."[323] The Mount Massive area is a good place to find more of the wild, and the Colorado Trail will take you partway there.

In *Colorado's Wilderness Areas*, John Fielder and Mark Pearson explain that the North American continent truly crests here, as the Rocky Mountains and the Continental Divide reach higher than anywhere else between the Arctic Ocean and the Isthmus of Panama.[324] Climbing lofty Mount Massive (on November 6) took me 11 hours (seven up and four down), partly because I spent about two hours unintentionally climbing 14,132-foot South Massive. Nearing its summit, I noticed several other climbers on a nearby peak, which was (literally) Massive.

Gerry Roach says, "Massive is massive. The mountain's name captures its essence. Massive has five summits above 14,000 feet on a 3-mile-long summit ridge. Massive is not just a peak; it is a region. No other single fourteener carries with it such a large area above tree line... No other peak in the 48 contiguous states has a greater area above 14,000 feet. By this measure, Massive reigns supreme."[325] Just below South Massive's summit, I took a late-morning lunch break, comfortably cocooned in the calming embrace, the enchanting trance of wilderness, and then regrouped and started toward my sixth official fourteener summit.

---

321 Backpacker magazine. "Rocky Mountain (Extra) High." *Backpacker*: October 2008, p. 20.

322 Backpacker magazine. "The Colorado Trail." *Backpacker*: January 2010, p. 19.

323 Susan Paul. "Wakening in the Wild: Discover The Colorado Trail with Sara Winter Nye." *PikesPique*: December 2006/January 2007, p. 1.

324 John Fielder and Mark Pearson. *Colorado's Wilderness Areas*. Englewood, CO: Westcliffe Publishers, 1994 p. 201.

325 Gerry Roach. *Colorado's Fourteeners*. Golden, CO: Fulcrum Publishing, 1999, p. 84.

After completing Mount Massive, I ventured south into the Sangre de Cristos to climb Humboldt Peak (14,064 ft.) during mid-November. "Sangre de Cristo" is, as previously mentioned, Spanish for "Blood of Christ," and many attribute the name to the shades of red that illuminate this range as night falls. The Sangre de Cristos start where the Sawatch Range ends, south of Salida, Colorado. Continuing south from there for 220 miles into New Mexico, the Sangres host some of Colorado's most rugged hikes and climbs.[326]

The Sangre de Cristos are *fault* block mountains, which differ from the *thrust* fault mountains covering the rest of the state. Fault block mountains rise up as one solid block, usually due to volcanic action. As the ancient volcanic San Juan Mountains (west of the Sangres) swelled from the valley floor, the earth tore apart, forming the Rio Grande Rift and driving the fault block of the Sangre de Cristos to their soaring heights.[327] In laymen's terms, the San Juans were formed through massive volcanic activity and the Sangre de Cristos by continental upheaval.

Since earning an MS in geology, Sheila Steele has spent much of her career teaching in Colorado. Among Colorado's unique features, she explains, is the Rio Grande Rift, which cuts through central Colorado and the western U.S. "It's an active feature, still moving, a split in the earth's crust that generates the mountain ranges—the valley that sinks and the mountains that rise," she says. "The zone is also crossed by our mineral belt, which is why Colorado has so many minerals—the world's largest molybdenum deposits and gold camps in Cripple Creek."[328]

The Sangre's Humboldt Peak was named after Alexander von Humboldt, a climber who gained fame in mountaineering circles after an unsuccessful attempt at climbing Ecuador's Mount Chimborazo in 1802, then thought to be the highest mountain in the world.[329] Mr. Humboldt's Colorado namesake rests less than two miles northeast of its better-known neighbors, Crestone Needle and Crestone Peak.

Gerry Roach says, "Humble Humboldt sits by itself 1.0 miles northeast of South Colony Lakes and 1.8 miles east of Crestone Peak... From Humboldt's summit and western slopes, there are superb views of the northeast face of Crestone Peak and Crestone Needle. Humboldt is a great place to either nervously preview or triumphantly review Crestone climbs. For those souls who have no intention of ever climbing Crestone Peak or Crestone Needle, Humboldt offers a safe vantage point in the heart of this exclusive place."[330]

---

326 http://www.summitpost.org/show/mountain_link.pl/mountain_id/298.

327 Saguache County 2008 Visitor's Guide (p.9): www.saguachecounty.net.

328 Lori Spaulding. "Geology teacher offers fun perspective of Colorado landscapes." *PikesPique*: October 2008, p. 1.

329 http://www.fourteenerworld.com/.

330 Gerry Roach. *Colorado's Fourteeners*. Golden, CO: Fulcrum Publishing, 1999, p. 162.

In contrast to the Class 3, 4, and 5 climbing common in the Crestone Group, Humboldt offers a more moderate Class 2 climb on its standard West Ridge route, and it's a perfect introduction to Colorado's fourteeners. Although not technically difficult, Humboldt requires a long approach (up to seven miles) and plenty of vertical (4,560 feet).[331]

Humboldt can be approached using several trailheads, but the most popular route starts near South Colony Lakes on the east side of the range, about 12 miles from the small town of Westcliffe. From there you could (until 2009) coax a 4-wheel-drive vehicle to within two miles of the mountain proper. On the other hand, South Colony Road was known as one of the worst in all of Colorado, and the Forest Service made a wise choice in closing part of it.[332]

Colorado Springs *Gazette* outdoors columnist R. Scott Rappold says South Colony Basin can get pretty crowded on summer weekends. "Tents cover the valley, and nearly every bush has been turned into a toilet by those who have flocked there to camp and climb the three fourteeners that tower over the frigid lakes." Mike Smith, a forester with the San Isabel National Forest, confirms Scott's observation. "When there are that many people, there are basically people... putting tents between their cars," Smith said.[333]

To manage the human impact on this fragile basin, the Forest Service closed the South Colony Road a couple miles below the site, building campsites and bathrooms and–to pay for maintenance–charging a $10 to $20 per-person fee.[334] The Colorado Mountain Club's Conservation Department did a poll of CMC trip leaders to gauge sentiment over the proposal. Several leaders opined similarly, saying: "The area has been trashed out for years, with people camping anywhere and disposing of human waste wherever they want. It's about time the Forest Service did something."[335] Here R. Scott Rappold provides a May 2010 South Colony Basin update:

> "The U.S. Forest Service Tuesday unveiled plans to begin charging hikers and backpackers a fee, $10 per person, per trip to hike and $20 to camp, in heavily used South Colony Basin in the Sangre de Cristo Mountains, an access point for four fourteeners. It would be the first permit and fee requirement on a fourteener, with the exception of Culebra Peak, which is privately owned ..."[336]

---

331 http://www.summitpost.org/show/mountain_link.pl/mountain_id/298.

332 http://www.summitpost.org/show/mountain_link.pl/mountain_id/298.

333 R. Scott Rappold. "Tent, matches, compass-wallet?" *The* [Colorado Springs] *Gazette:* 4/2/08.

334 R. Scott Rappold. "Tent, matches, compass-wallet?" *The* [Colorado Springs] *Gazette:* 4/2/08.

335 Steve Bonowski. "Recreational access." *Trail & Timberline:* Fall 2006, p. 13.

336 R. Scott Rappold. "Days of free fourteener climbing may be ending." *The* [Colorado Springs] *Gazette:* 5/12/10.

"Near Westcliffe, the basin near timberline is the closest fourteener trailhead to Colorado Springs, after Pikes Peak. Humboldt Peak, Crestone Peak, Crestone Needle, and Kit Carson Peak can all be climbed from there, and officials say up to 4,500 people visit annually, in an area with no restrooms. Last fall, the Forest Service banned campfires near the lakes, closed South Colony Road to vehicles 2.6 miles below and built campsites at the closure, where the $20 overnight fee would also apply."[337]

Colorado has far too many roads and ATV trails crisscrossing and degrading its national forests and other public lands anyway. There are 14.5 million acres of Forest Service land in Colorado, with most open to mixed use, including off-road vehicles and energy exploration.[338] As a result, today, only 8 percent of the national forest acreage in Colorado lies beyond one mile of a road (only 4 percent for BLM lands), and there are enough Forest Service roads in the state to go from the Kansas border to Utah and back, 17 times.[339]

Spencer Swanger, the first person to scale the state's 100 highest peaks, said even back in the late 1970s, wildness and solitude were getting hard to come by in Colorado, which is why he sought out the far corners. "Most of the Colorado wilderness is pretty well dissected by roads," he told The [Colorado Springs] Gazette after completing the 100 peaks. "It's hard to think of someplace where you'll get on top and don't see a road or town."[340] Hence, we can certainly get by with one less (thousands less, actually) road on our national forest, BLM, and other public lands.

Colorado's foremost hunter-conservationist, David Petersen, thinks so too. "The futures of hunting, fishing, outfitting and other backcountry activities, together with the significant benefits they bring to our state's celebrity, dignity and economic welfare, depend on keeping Colorado's last remaining islands of unspoiled public wildlands intact."[341] But closing habitat-damaging roads or motorized trails in Colorado or anywhere else tends to be problematic, as explained by Salida, Colorado, writer Ed Quillen.

"The way things generally work around here," says Ed, is "the [no motorized vehicles] sign post will be knocked over or pulled out within a week or two. The blocking rocks will be pushed away so that motorcycles and smaller ATVs can get through. The BLM will likely return with better blockage, but meanwhile the motorheads will be writing letters to the local paper, complaining about how

337 R. Scott Rappold. "Days of free fourteener climbing may be ending." The [Colorado Springs] Gazette: 5/12/10.
338 Bill McKeown. "Roadless areas touted as critical to Colorado." The [Colorado Springs] Gazette: 1/5/06.
339 Alan Kesselheim. "Lewis & Clark's Wild, Wild West." Backpacker: February 2003, p.38.
340 Scott Rappold. "Local climbing legend dies in fall." The [Colorado Springs] Gazette: 7/22/10.
341 David Petersen. "Sportsmen applaud Ritter's roadless rule action." Aspen Daily News: 12/20/08.

they've been 'locked out of our public lands,' and that it's elitist to restrict access to people with $40 shoes rather than $6,000 ATVs. Plus some old-timer learned to drive there with a 1937 Studebaker pickup, and closing the road would deprive him of his custom, culture and heritage."[342]

The South Colony Lakes road approach is a fairly nasty drive or moderate hike, but the mountain panorama at its terminus includes three magnificent fourteeners–Humboldt Peak, Crestone Peak, and Crestone Needle. *National Geographic Adventure* contributor Gabby Anstey says, "South Colony will turn your vehicle into a lunar rover. If you're skilled in pitted, precipitous terrain, go for it. Otherwise, leave your ride... grab your camping gear and supplies for the night, and walk the four miles to the trailhead, then another mile or so to the South Colony Lakes."[343]

I opted to drive the South Colony Road, then car-camped at the trailhead and started up Humboldt Peak at first light, clearing tree line in time to see two bighorn sheep sparing in the early-morning sun. They were on a windswept ridge far above, but soon noticed my approach and disappeared into the piles and miles of brown boulders, massive slabs, and crumbly cliffs that are Humboldt Peak. Encountering high-mountain inhabitants like bighorns, mountain goats, elk, mule deer, marmots, ptarmigan, pika, etc., is one of my main motivations for climbing 14ers, in addition to the unbelievable beauty and solitude.

> "Being in the wild like that, you're really able to take
> a step back... and experience the pureness and
> goodness of what I think God created."
> –Michael Crotteau[344]

I climbed a total of eight fourteeners during 1999. Humboldt Peak was number seven (on November 13), followed by Mount Bierstadt on November 20. Bierstadt is not a difficult hike-climb, but like all 14ers, is potentially dangerous, especially if you're unprepared or unwary. Along with Mount Evans, Grays Peak, and Torreys Peak, Bierstadt is one of the closest fourteeners to Denver, and it's the 38th highest peak in Colorado.

The most popular trailhead is located on the Guanella Pass Road, only an hour-plus drive from the Denver metropolitan area. The Guanella Pass Scenic and Historic Byway, a 22-mile dirt roadway, climbs north from Grant (on U.S. 285) up to 11,669 feet at the pass, and then descends to Georgetown and I-70, winding <u>along between</u> the Mount Evans Wilderness and Square Top/Burning Bear road-

342 Ed Quillen. "Thank you, BLM." *High Country News*: 3/18/09.
343 Gabby Anstey. "Cruising Back-Road Colorado." *National Geographic Adventure*: August 2003, p. 66.
344 Christine Schmid. "Juneau adventurer paddles through Yukon River history." *Anchorage Daily News*: 11/2/03.

less areas.[345] Like many mountain roads, this one started out as a wagon trail, connecting the gold and silver mining towns of Georgetown and Grant.[346]

In his book *Battle for the Wilderness*, Michael Frome says Mount Bierstadt was named after nineteenth-century adventurer and painter Albert Bierstadt. Bierstadt was the first prominent landscape artist to head west, joining a military expedition to the Rocky Mountains, in 1859. He sketched and painted everywhere he went, and looking up at the mountain that bears his name, it's obvious why he came to Colorado, and likely thoroughly enjoyed his chosen profession. [347]

From Guanella Pass, you can see Mount Bierstadt's 14,060-foot summit flanked by a distinctive jagged lower peak known as "The Sawtooth." The standard Bierstadt route (the Western Slope) is only a three-mile-long, 2,391-foot slog up a Class 1 trail, but for those interested in more of a challenge, there are (as always) more difficult routes. One of them is the Class 3 Sawtooth Ridge, which connects Mount Bierstadt and Mount Evans.[348] Bierstadt sits just 1.4 miles west of Evans, but I opted for the more direct Western Slope route on this, my maiden hike-climb of Bierstadt.

Although there wasn't a cloud in the sky on this late November morning, strong winds near the summit made for bone-chilling temperatures that threatened to freeze exposed skin in seconds. I was properly dressed for the weather, but quickly gained a deep respect for the Colorado Rocky Mountains and their unforgiving temperament. Not surprisingly, there were only two others tempting a frozen fate on the summit with me during this frigid morning.

Since then, I've read more than a few stories about unwary Denverites and others going up Bierstadt for a leisurely afternoon stroll, ill-equipped for nasty weather, then getting pinned down by the gale-force winds that frequently whistle over Guanella Pass. I've also seen perhaps hundreds of hikers now venturing above tree line in pursuit of 14ers with little more than cotton shorts and light T-shirts for protection. These are the folks who often get into trouble and, in the worst cases, end up dead or seriously injured, endangering the lives of those who have to save them or retrieve their bodies. This 2001 *Denver Post* story is reflective of what can happen to the unlucky or unprepared.

"Three members of a Denver-area family were rescued Monday after spending a harrowing night atop 11,669-foot Guanella Pass in a spring snowstorm. Richard McQueen [who was wearing blue jeans], 50, his son, Brad McQueen, 27, and his daughter-in-law, Melissa McQueen, 25,

345 Tod Bacigalupi. "Guanella Pass Disrupts Wildways While US 285 Restores Connectivity." *Landscapes*: May 2009, p. 8.
346 Rick Spitzer. "Easy access to fourteeners." *Vail Daily*: 9/12/09.
347 Michael Frome. *Battle for the Wilderness*. Salt Lake City: The University of Utah Press, 1974, p. 38.
348 http://www.summitpost.org/show/mountain_link.pl/mountain_id/10.

were returning from a hike up 14,262-foot Mount Evans when Sunday's storm blew in with blizzardlike conditions and stranded them a half-mile from their car.[349]

"The three sought shelter amid trees and bushes, rescuers said, but suffered frostbite and hypothermia. 'They weren't really prepared for the drastic change in the weather,' said Clear Creek County Sheriff Don Krueger... The storm... dumped six inches of wet snow on the area, produced whiteout conditions, and dropped temperatures well below freezing... 'It ought to be a good reminder that you never go out in the high country in Colorado without being prepared...,' Sheriff Krueger said."[350]

For the cotton-wearing crowd, here's some more sound advice from northern Minnesota wilderness traveler (and paddler) Doug Smith: "A pair of blue jeans may be comfortable in the backyard, but [in the mountains]... they can be trouble. Get them wet, and they may be damp for days. Use mostly synthetic clothing such as nylon and polypropylene that dries quickly. I was caught in the middle of a lake when a rainstorm struck... drenching me. But my pants and shirt were bone-dry in a matter of minutes after the rogue cloud departed. Leave cotton at home. Fleece jackets retain warmth even when wet. And don't forget good rain gear."[351]

Too many hikers get into trouble on Bierstadt and other 14ers primarily because so many of them are completely unprepared for anything but bluebird weather. According to *The Denver Post*, "There are so many people–from young whippersnappers to truly elderly–hiking on weekends, it looks like a trip to the mall."[352] Having spent much of my life in the outdoors, I had enough commonsense to be prepared with both the appropriate gear and mind-set. Mountaineering and climbing, like many of life's endeavors, are (in my opinion) roughly 50% commonsense, 40% experience, and 10% luck. If you lack commonsense, experience and luck may not do you much good.

As Kurt Tucholsky said, "There's no such thing as experience. A man can do a thing badly all his life."[353] However, you can never disregard experience, as my friend Gary Scott will tell you: "Mountaineering is a dangerous game. I don't count anymore, but I know I've lost well over 30 friends in the mountain ranges of the world, and I really don't want to lose any more... Many people are VERY lucky in the mountains. Experience is what keeps you alive. Years and years of

349 Marilyn Robinson. "3 survive frigid night atop pass." *The Denver Post*: 8/26/01.

350 Marilyn Robinson. "3 survive frigid night atop pass." *The Denver Post*: 8/26/01.

351 Doug Smith. "Boundary Waters 101." [Minneapolis-St. Paul] *Star Tribune*: 4/27/10.

352 The Denver Post. "Top 10 things to do in Colorado before you die." *The Denver Post*: 7/14/10.

353 Reinhold Messner. *All Fourteen 8,000ers*. Seattle, WA: The Mountaineers, 1999, p. 15.

rock, snow, and ice in all conditions gets you that experience."[354]

Colorado's Fourteeners author Gerry Roach attributes his climbing successes to adhering to "the fundamentals" and knows that rock climbing is "the basis of all mountaineering." "I got well schooled in that before I ever went to the mountains," he says. "I'd practice the basics—rock climbing, snow climbing, self-arrest, setting anchors and belays, crevasse rescue—and I kept those skills topped up."[355]

Teton (Wyoming) Range guide Jack Turner knows from his "fundamental" climbing experience not to take chances in the mountains. "Mountains have many moods," he says. "Even under clear summer skies I require my clients to pack warm clothing, to be prepared for the worst. I am a climbing guide, and like all guides, I am a skeptic about mountain weather. We abide by a local adage: Only fools and newcomers predict the weather in the Tetons [or Rockies]."[356] Mountain weather, like life in general, is inherently unpredictable. No matter how clear the skies or calm the winds, a peaceful, serene setting can turn ugly in an instant.

Climbing and mountaineering are, at their best, fitting metaphors for life and making a difference. At their worst, blind obsessions that can, and often do, kill the most experienced climbers in the world. On the other hand, all truly motivated and successful climbers and others are obsessed with their pursuits to a point. As my South African friend Ronnie Muhl says, "On one level I knew that obsession can lead to irrational behaviour, but on another level I knew that if correctly channeled, it could be an unstoppable force."[357]

In "Confessions of a Solo Climber," *Outside* contributor Mark Jenkins confesses some of his sins: "Surviving by the skin of your teeth is the stuff of legend. These are the war stories we boast of, as if survival were vindication. But it's not. Just because I summited and got down alive doesn't mean I did the right thing. I'd been a fool. I'd made bad decisions... Surviving after a series of stupid moves is nothing more than the Goddess of Good Fortune taking pity on you. (Don't ask her to do it more than once.) It's nothing to be proud of."[358]

Eleven-year-old fourteener completer Anthony Wada's logic is indisputable on this point: "Mother Nature is not always nice. You don't have to climb 14ers if you don't want to."[359] Whether it's slipping and taking a deadly fall, succumbing slowly to hypothermia, freezing to death in a lashing storm, being buried by <u>an avalanche,</u> or struck by lightning, Mother Nature and the mountains don't

354 Gary Scott. "Another friend bites the dust!" E-mail: 9/13/08.

355 Lori Spaulding. "Roach finds adventure on peaks near and far." *PikesPique*: March 2010, p. 1.

356 Jack Turner. *The Abstract Wild*. Tucson, AZ: The University of Arizona Press, 1996, p. 19.

357 Ronnie Muhl. *Everest: Surviving the Death Zone*. Inspiration at Work Publishing: Cape Town, South Africa, 2008, p. 137.

358 Mark Jenkins. "Confessions of a Solo Climber." *Outside*: February 2000, p. 37.

359 Susan Baker. "Father-son team complete fourteeners." *Trail & Timberline*: Winter 2007-2008, p. 37.

take pity on anyone or anything. In the long run, only a solid dose of commonsense and caution combined with carefully calculated risk-taking, and some luck, will get you onto the summits of all the 14ers without experiencing at least a minor injury or two.

In "Peak experiences," *Anchorage Daily News* contributor Melissa DeVaughn says the best climbers merge their physical abilities with their mental capacity to read and reduce dangers. They know when to turn around—even if it means being only a few hundred feet from the summit. They know when to press on—even when their bodies are feeling the stress. "There is always danger, but you never think about that—at least I don't, because of the planning," says Jim Donini (president of the American Alpine Club). "I'm always making the danger minimal."[360]

For those who get "Everest fever" and continue climbing when commonsense and instinctual caution are telling them to turn around, climber Greg Child has a sobering message:

> "There is a state of mind that sometimes infects climbers in which the end result achieves a significance beyond anything that the future may hold. For a few minutes or hours, one casts aside all that has been previously held as worth living for, and focuses on one risky move or stretch of ground that becomes the only thing that has ever mattered. This state of mind is what is both fantastic and reckless about the game. Since everything is at stake in these moments, one had better be sure to recognize them and have no illusions about what lies on the other side of luck."

Even more to the point, Herman Buhl says, "Mountains have a way of dealing with overconfidence." Greg's and Herman's words of wisdom and warning lead us, again, to the question often asked of mountaineers and other adventurers. Why? Given all the potential though unlikely dangers associated with tackling 14ers, why do it? Why spend hours or days and weeks hiking and climbing up and down mountains? Why endure all the physical exhaustion and pain? Why bother? Stacy Allison, the first American woman to climb Mount Everest, knows why:

> "I climb because I'm here. I don't battle the mountains, I don't conquer anything, even when I do pull myself onto a summit. For me, the triumph comes in every step, in every breath and heartbeat along the way. It's the sheer pleasure of being on the planet, seeing the mountains around me and, for a brief moment, being a part of them. I climb for a simple reason: because I'm alive."

---

360 Melissa DeVaughn. "Peak experiences." *Anchorage Daily News*: 11/11/07.

In questioning his motives for climbing Mount Everest a second time, Ronnie Muhl wrote: "In contemplating this question, I remembered what Dr. Jim Litch, who had been to [Mt. Everest] Base Camp many times, had to say: 'Life in the mountains draws out the character of those who journey there.' He went on to write: 'Maybe this is one of the reasons we climb—to see ourselves at the core, not packaged and contained as we are when living within the constraints of technology and consumerism.'"[361]

Ronnie continues, saying: "Nick Heil, the author of *Dark Summit*, answers this question for me in a most profound way. He says, 'What folly to think that anyone climbs Everest for the views, or the thrills, or the bragging rights, or, vaguest of all, because it's there. What's there is this: the chance to be worthy of your own dreams.' I think we should constantly be challenging ourselves by asking *Why?*"[362] Me too.

The greatest gift of the democratic freedoms we enjoy in this country is the God-given (possibly) and constitution-given (verifiably) right to make our own decisions; our individual right to live as we wish, to come and go as we desire, to believe or not believe as we individually choose, and to dream. I choose to climb mountains, to travel the world, and explore its wildest places in order to experience the *real* world; the one that gave birth to our species, and the one we must save if we are to save ourselves.

I do it in order to live life fully and to use those experiences to try and make a difference in a world that desperately needs our immediate attention. I do it in order to enlarge my perspective on life and learning and to understand what we're losing as the earth's remaining wild places are diminished and developed; I do it to gain a better understanding of what we must do to save them. World-renowned climbers Reinhold Messner and George Mallory know why.

According to Messner, "This state of merging into infinity is a sensation I have frequently experienced on big mountains, and it always seems to accentuate the existential problems of man. Why are we here? Where do we come from? Where are we going? I have not discovered any answers, and if you discount religion, there are no answers, only that the state of being active within life activates the fundamental questions of Life. Up there, I didn't question what I was doing, why I was there. The climbing, the concentration, the struggle to push myself forward, those were the answers."[363]

Mallory adds, "The first question which you will ask and which I must try to answer is this, 'What is the use of climbing Mount Everest?' And my answer must <u>at once be, 'It</u> is no use.' There is not the slightest prospect of any gain whatso-

---

361 Ronnie Muhl. *Newsletter 257:* 2/9/10.

362 Ronnie Muhl. *Newsletter 257:* 2/9/10.

363 Reinhold Messner. *All Fourteen 8,000ers.* Seattle, WA: The Mountaineers, 1999, p. 30.

ever. Oh, we may learn a little about the behavior of the human body at high altitudes, and possibly medical men may turn our observation to some account for the purposes of aviation. But otherwise nothing will come of it.

"We shall not bring back a single bit of gold or silver, not a gem, nor any coal or iron. We shall not find a single foot of earth that can be planted with crops to raise food. It's no use. So, if you cannot understand that there is some-thing in man which responds to the challenge of this mountain and goes out to meet it, that the struggle is the struggle of life itself upward and forever upward, then you won't see why we go."[364]

Climbing and hiking, trekking and traveling, experiencing the world in general, all serve to enlarge our perspective on life, learning, and making a difference. Northern Minnesota writer Jim Dale Vickery says they "teach us skill at comparison and broaden our parameters of understanding. They enable us to clarify, define, and love the homes and lands we came from. They foster appreciation of geo-graphical roots; their waters, mountains, plains, skies, climates, plants, and animals, each in its turn becoming more distinct in the illumination of other places."[365]

Ultimately, Vickery says, "the goal is to comprehend life's greater mosaic. It is, in the end, a balancing act: a juggling of space and time for maximum view." T. S. Eliot adds, "The end of all exploring will be to arrive where we started, and know the place for the first time."[366] But one must travel and explore, not infrequently, in some way, shape, or form in order to attain such enlightenment. There is no substitute.

I hike and climb mountains for many reasons (wildlife viewing, spiritual re-newal, solitude, perspective and understanding, exercise, to be worthy of my dreams, etc.), but maybe the most important reason is that when I'm old and gray, sitting in my driveway in a lawn chair, I won't be one of those brooding, crestfallen old men who has to spend the few remaining years of his life regret-ting adventures he never had, dwelling until death on what could have and should have been.

Regret. It's one of our strongest emotions and most potent motivators. I re-gret, like us all, many things I've done and have not done during my life. I regret not always treating others with the dignity and respect they deserve. I regret that I am not closer (through no fault of hers) to my sister, my only sibling. I regret the loss of our parents. I regret that my solitary pursuits and passions may have desensitized me to the importance of family and to starting a small one (no more than one or two children) of my own.

---

364 George Mallory. "Mount Everest Quotes." *www.mnteverest.net.*
365 Jim Dale Vickery. *Open Spaces.* Minocqua, WI: NorthWord Press, Inc., 1991, p. 247.
366 Ronnie Muhl. *Everest: Surviving the Death Zone.* Inspiration at Work Publishing: Cape Town, South Africa, 2008, p. 190.

But for each regret, I have a hundred enlightening life experiences and thousands of priceless memories gleaned from years of hiking, climbing, hunting, traveling, reading, writing, and learning. And I have decided that no matter what, I will not someday look back on my life and regret that I have not lived it fully, not endeavored to make a difference, not recognized the importance of the natural and wild world to humanity's continued survival and evolution.

"A civilized society exhibits five crucial characteristics," the British philosopher Alfred North Whitehead said, "peace, art, beauty, truth, and adventure. Without adventure, civilization is in full decay." Antarctic explorer Apsley Cherry-Garrard adds, "Exploration is the physical expression of the Intellectual Passion."[367] Helen Keller seemed to agree: "Security is mostly superstition. It does not exist in nature, nor do the children of humans as a whole experience it. Avoiding danger is no safer in the long run than outright exposure. Life is either a daring adventure or nothing."[368]

Taking Alfred's, Apsley's, and Helen's advice to heart, I will not grow old and regret not having done good and adventurous things with my life. I won't regret what should have or could have been, but instead will rejoice and reminisce in the slowly fading glow, in the fond memories of a life full of exploration and adventure, of one well-lived.

I have strived to live so that someday when I'm on a plane or mountainside plunging headlong toward earth and an all-but-certain death, I will not have gone gently into that good night, as Dylan Thomas extolled, but will hopefully have been an example of what is possible, right, and just in this world. If each of us does the same, we can truly move mountains and make the world a better place for future generations. That's why I climb. That's why I endeavor to make a difference.

> "I went to the woods because I wished to live deliberately,
> to front only the essential facts of life, and see if I could
> learn what it had to teach, and not, when I came to
> die, discover that I had not lived."
> —Henry David Thoreau

---

367 Apsley Cherry-Garrard. *The Worst Journey In The World*. New York: Carroll & Graf Publishers, Inc., 1922, p. 597.
368 Ronnie Muhl. *Everest: Surviving the Death Zone*. Inspiration at Work Publishing: Cape Town, South Africa, 2008, p. 252.

1999's Fourteeners (8):

Colorado: Pikes Peak, the Gold Camp Road, Mueller State Park, Castlewood Canyon State Park, the Crags, and Florissant Fossil Beds National Monument (29 May - 31 May)

Colorado: Saint Mary's Falls Trail, Helen Hunt Falls, and the Gold Camp Road (6/5/99)

Colorado: Rocky Mountain National Park (11 June - 13 June)

Colorado: Rocky Mountain National Park and Mount Evans (2 July - 4 July)

Colorado: Windy Ridge Bristlecone Pine Scenic Area (7/31/99)

1. Longs Peak (8/14/99)[369]

2. Mount Elbert (8/28/99)[370]

3. Grays Peak (9/4/99)

4. Torreys Peak (9/4/99)

La Plata Peak (9/11/99)[371]

---

369 I completed the 15-mile (round-trip) Longs Peak Keyhole route in 14 hours. See article: David A. Lien. "Longs Peak: First Fourteener." *Trail & Timberline*: Winter 2006-2007, p. 12.

370 Mount Elbert (14,433 ft.) is the second-highest point in the Lower 48 states, the highest in Colorado and the entire Rocky Mountains, and was my second state highpoint.

371 I turned back short of La Plata Peak's summit due to weather (snow and thunder combo).

5. La Plata Peak (9/25/99)[372]

Utah: Canyonlands National Park-Horseshoe Canyon Unit and The Maze District, Goosenecks State Park, and Hovenweep National Monument (2 Oct. - 8 Oct.)

Kansas: Geographic Center of the United States (10/21/99)[373]

6. Mount Massive (11/6/99)

7. Humboldt Peak (11/13/99)

8. Mount Bierstadt (11/20/99)

---

372 There was solid snow cover above tree line on La Plata Peak, and knee-deep snow near the summit. I was the first person to reach the summit that morning.

373 The Geographic Center of the U.S. is near Lebanon, Kansas.

"To me, trekking is a very good approximation of heaven on earth. I love the sense of going somewhere, of a journey to a destination rather than a mere aimless wander. I love the remote landscapes and the local people going about their regular business. I love the walking, the mindless putting of one foot in front of the other. I love the utter distance from my normal life."
–David Noland[374]

"Why was I getting so worked up over some episode in Washington or development on the political trail that turned out to be an asterisk in the long course of human events? When you get out *there*, into the wilderness, you see all around you the long course of time. It puts you in your place, and that's very useful."
–Tom Brokaw[375]

---

374 David Noland. *Outside*: March 2001, p. 60.
375 Tom Brokaw. "Dispatches: Anchor's Away." *Outside*: December 2004, p. 30.

"Distance changes utterly when you take the world on foot. A mile becomes a long way, two miles literally considerable, ten miles whopping, fifty miles at the very limits of conception. The world, you realize, is enormous in a way that only you and a small community of hikers know. Planetary scale is your little secret. "
–Bill Bryson, *A Walk in the Woods*[376]

"One of the problems of modern times is that we are separated from the world that supports us by the speed with which we traverse it. Walking is the best way to know a place, perhaps the only way."
–Chris Townsend, *Walking the Yukon*[377]

---

376 Bill Bryson. *A Walk in the Woods*. New York: Broadway Books, 1998, p. 71.

377 Jan Adkins (ed.). *The Ragged Mountain Portable Wilderness Anthology*. Camden, Maine: Ragged Mountain Press, 1993, p. 5.

Climber standing in the Keyhole (13,160 ft.) on Longs Peak: 14 Aug. 1999

David on the summit of Longs Peak (14,255 ft.): 14 Aug. 1999

Helicopter circling to retrieve body of fallen climber at the Ledges on Longs Peak: 14 Aug. 1999

Helicopter with body of fallen climber viewed from the Ledges on Longs Peak: 14 Aug. 1999

2000: Peak Year

Edward Abbey summited a few mountains in his day and wrote about some of his climbing experiences in *The Journey Home*: "I find the first step upward so difficult that each time I begin the ascent of a mountain I swear to myself, Never again. I say, and mean it, This is the last time. The body objects, the heart and lungs complain, and gravity, with arms of lead, drags at our limbs, pulls down our vanity.[378]

"And yet the pain of it all is soon forgotten (like childbirth, they say), and a week or a month or a half-year later," Ed says, "we're at it again, trudging with iron shoes and pig-iron on the back up yet another mountain trail, toward one more ugly, meaningless, and brutal rockpile in the sky. If we did it for pay, we'd call it slave labor. What punishment could be so cruel and unusual as that which is self-inflicted?"[379]

There were few indications that my pre-Colorado outdoors experiences would eventually lead to the "self-inflicted punishment" of climbing fourteeners, and there was a time when I may have agreed with Ed's lighthearted climbing condemnation, but I grew up hunting, hiking, camping, trapping, and canoeing amidst the woods, wilderness, lakes, streams, and rivers of northern Minnesota. Like many young boys in that part of the country set loose to roam the woods after school and on weekends, I lived for hunting, fishing, camping, and trapping, and made little or no distinction between these activities and life at large.

Conversations with friends at school were generally mixed with talk of hunting deer, ducks, geese, grouse, and woodcock; trapping mink, muskrat, and beaver; pumps vs. automatics; and hunting cabins and shacks. During those years and later in life, I did an awful lot of hunting-related scouting, exploring, and bushwhacking, and I even hiked a few peaks before moving to Colorado.

378 Edward Abbey. *The Journey Home*. New York: Penguin Books, 1991, p. 214.
379 Edward Abbey. *The Journey Home*. New York: Penguin Books, 1991, p. 214.

In retrospect, it may have been inevitable that I'd eventually set my sights on 14ers.

Hunting, in particular, provided invaluable preparation for the rigors of mountaineering. Years of stalking big and small game during freezing cold mornings, snowy afternoons, and steel-gray evenings introduced me to the many moods and extremes of the wilds. Hours filled with silent thought and quiet reflection, miles of slow plodding and methodical scanning, senses alert and focused, always prepared for the unexpected, the unlikely. Immersed in this untamed realm of predator vs. prey, I learned the potential dangers of wilderness travel and the basics of mountain survival.[380]

A deer's silent approach and crashing through the underbrush retreat; a grouse's explosion from underfoot and wild waylay through the aspens; a woodcock's graceful, almost vertical liftoff and fanciful flight for safety; a mallard's low, fast flyby and unscathed escape. A mishandled rifle and accidental but deadly discharge; a bludgeoning rockfall or roaring avalanche; an instantaneous flash-boom lightning strike; a misplaced step and slip into oblivion. In this ephemeral state, balanced sometimes precariously between life and death, hunting and climbing are one.

In northern Minnesota, the fall deer hunting season is regarded as practically sacred, and whitetails are found statewide in abundance. Just ask Governor Tim Pawlenty. One day as Pawlenty arrived for work at the Capitol in St. Paul, he and his staff heard shattering glass, then saw a buck charge past five feet away. The deer broke two windows before bounding off. During mating season, rutting bucks sometimes charge their own reflections. Pawlenty said the incident was a good omen: "We could have had the governor's deer opener right here at the Capitol."[381]

Yes, Governor Pawlenty is a deer hunter. "Deer hunting is a part of our tradition and heritage in Minnesota," he says. "It allows people to come together who are friends, who are neighbors, who are family members, co-workers, and they spend time together in the great outdoors in Greater Minnesota... they spend time swapping stories, exchanging ideas, telling tales and just being friends–being good dads and moms, sons and daughters, grandmas and grandpas. It builds that sense of friendship and community and family, and that's a good thing."[382]

While growing up in Grand Rapids, Minnesota, my friends and I often took time off from school during the fall to hunt deer, whereas any other time of the

---

380 David A. Lien. "Hunting & Climbing: Age-Old Quests." *Whitetales*: Summer 2006, p. 32.

381 Associated Press. "Deer leaves its mark on state Capitol." *Duluth News Tribune*: 11/4/05.

382 Warren Wolfe. "Gov. Pawlenty fails to bag deer, but uses day to talk up hunting." [Minneapolis-St. Paul] *Star Tribune*: 11/9/03.

year, a death in the family or a doctor's note would have been the only acceptable excuses. Deer hunting was different though. It isn't just a "sport" to me and
most others who hunt. A sport is something you do for entertainment. Hunting
and shooting animals for food is not entertainment. In fact, as BHA member
Dan Crockett says, "Perhaps there are aspects of our relationship with a wild
animal that can only be had by hunting it."

Dan knows that those who hunt not only acknowledge, but accept personal
responsibility for the reality that we all kill, that every creature exists at a cost to
other creatures; life is built upon death.[383] According to Dan, "Choosing to hunt
raises no one to higher ground. It merely opens a pathway into a different land.
This is a magical place where insights into the spirits of both the hunted and the
hunter may be revealed. It is a land forever tinged with sorrow."[384]

As Dan and other hunters know well, hunting is about much more than the
kill. It's about getting away from cities and civilization and reconnecting with
nature and creation, it's about participating fully in the natural cycle of life and
death, about experiencing the real (wild) world, and escaping the one we've
created from it. One might easily use the word religion here; a feeling of con-
nection to the land, the animals, and something far larger and more important
than ourselves.

Some will surely scoff at such sentiment, but hunting is, without a doubt, one
of the most philosophical endeavors in which humans can engage. Life and
death struggles ensue. A deer may be taken by the hunter or it may escape
and survive. The hunter may become lost, freeze to death, fall from a stand,
or be mistaken for the quarry. Such are the challenges and risks during the
chase. Our prehistoric ancestors faced similar dangers as they gouged with
spears upon hairy mastodons and were, if not careful, trampled underfoot by the
charging beasts.

Today, hunting is still a deadly and messy business, make no mistake. Edward
Abbey said it's one of the hardest things to even think about: "Such a storm of
conflicting emotions!" Hunting is also one of the few activities that allows modern humans to participate directly in the life-and-death cycles upon which all
natural systems depend. You can buy your meat at the store and abhor those of
us who hunt, but all you've done is paid someone to do the killing for you.

Craig Medred (outdoors columnist for the *Anchorage Daily News*) reminds us
that death is a constant throughout the wilds of North America. Only in cities do
we think of it as somehow foreign, "Only in the cities do people get themselves
all in a tizzy that in the age-old struggle between predator and prey, man might

---

383 David Petersen. *Elkheart*. Boulder, CO: Johnson Books, 1998, p. xii.
384 David Petersen. *Elkheart*. Boulder, CO: Johnson Books, 1998, p. xiii.

play the role of hunter. Never mind that humans have been predators in North America since time immemorial. For most of our history, we had no choice. Back in the day, journeying to the supermarket was not a survival option."[385]

However, today, the supermarket is where most people do their hunting. The number of hunting license holders in the United States in 2004 was 14,779,071, which was up slightly from the 2003 total of 14,740,188. Sounds like a lot hunters. Problem is we currently represent a mere 5 percent of the U.S. population. Our numbers peaked most recently in the mid-1980s at 16.8 million, when we represented just over 9 percent.[386]

> "I think that our intelligence itself was largely
> evolved from being a hunting animal. That
> connection shouldn't be let go of."
> –Doug Peacock[387]

Today's hunters are a direct link to our hunter-gatherer past and a vital offset to the multiplying masses of urban dwellers who don't have time for, interest in, or access to the outdoors, and whose children are increasingly raised on video games and television. Hunters are among the most astute observers of nature. Up before dawn and keen to the subtlest cues in wildlife behavior, we represent an evolutionary trait and tradition that has propelled humanity through the distant and dangerous ages to an even more precarious present.[388]

As explained by Joel Achenbach, in "The Drive-Thru Wilderness," modern persons (nonhunters in particular) tend to live a life out of balance with their spiritual needs. Many of us, hunters and nonhunters alike, sit in offices working in front of computers. We are "appendages of a vast technological machine." Notice, by the way (says Joel), that to a remarkable degree our offices, workstations, computer monitors, and keyboards do not appear in our dreams. In our dreams, we go home, to the wilds, the woods, the mountains and deserts. In dreams, we are outside, unconfined and free.[389]

Northwoods photographer Jim Brandenburg dreams of wandering the woods like he did as a boy. "I felt... compelled to let go of life's clutter and a world lit by computer screens instead of the sun. I wanted to wander the forest again, to see what was over the next rise, to follow animal tracks in the snow with the eyes

---

385 Craig Medred. "To Hillside moose, hunters are just another predator." *Anchorage Daily News*: 11/6/05.

386 Sam Cook. "Field Reports." *Duluth News Tribune*: 11/6/05.

387 Ted Chamberlain. "Doug Peacock: Veteran of the Grizzly Wars." *National Geographic Adventure*: July/August 2000.

388 Steve Williams. "An Important Tool for Conservation." *Wild Earth*: Winter 2003-2004, p. 60.

389 Joel Achenbach. "The Drive-Thru Wilderness." *Wilderness*: 2004-2005, p. 41.

of a boy."[390] As do many of us, still, so the traditional rite of autumn for over 14 million Americans still goes on despite decreasing numbers of hunters and fewer wild places to hunt.

In his seminal work *A Sand County Almanac*, Aldo Leopold wrote: "Hunting in most of its forms maintains a valuable element in the cultural heritage of all peoples." Especially for those of us who live in places like my boyhood home of northern Minnesota, places that still have vast swaths of unbroken forests and wide-open spaces, places where wildlands and wildlife still hold their own against development and desecration.

Long hours, days, and weeks afield, straining to maintain senses and sometimes sanity, not knowing when or where you might be required to act but knowing split-second decisions will make the difference, that is hunting and climbing. A full freezer or an empty one, a summit or not, it doesn't matter. The experience, the transcendent exposure to nature and creation, is what matters. Solitude without loneliness, companionship without company, rhapsody and reverence without repentance. All these experiences are why I hunt, and they almost certainly predisposed me to climb one day.

"Just ask yourself, 'Why do kids climb trees?'" asks Jeff Achey, editor of *Climbing* magazine. "When you're moving up, whether in a yard or a gym or on a mountain, there's something primordial going on. And most of us never outgrow that."[391] In our not so distant past, climbing provided the hunted with places safe from predators, and the hunter with perches to search the surrounding terrain for prey. It's easy to forget in our seemingly modern age that humans evolved as both predators and prey.

Up until about 13,000 years ago, all human societies were based on the hunter-gatherer model. Edward Abbey felt we were better off then. "I believe humanity made a serious mistake when our ancestors gave up the hunting and gathering life for agriculture and towns," Ed wrote. "That's when they invented the slave, the serf, the master, the commissar, the bureaucrat, the capitalist, and the five-star general. Wasn't it farming that made a murderer out of Cain? Nothing but trouble and grief ever since."[392]

Renowned hunter-conservationist Theodore Roosevelt was a proponent of living close to the land too: "Over-sentimentality, over-softness, in fact washiness and mushiness are the great danger of this age and of this people. Unless we keep the barbarian virtue, gaining the civilized ones will be of little avail."[393] Hiking, backpacking, climbing, and mountaineering are all great ways to expe-

390 Jim Brandenburg. "Northwoods Journal." *Land & People*: Fall 2000, p. 37.
391 Marco R. Della Cava. "Teen scales heights of rock climbing." *USA Today*: 11/18/03, p.2D.
392 Edward Abbey. *Abbey's Road*. New York: Penguin Group, 1979, p. 141.
393 Frank Miniter. "The Ultimate Man's Survival Guide." *American Hunter*: June 2009, p. 24.

rience the outdoors, but I have a hunter's (and trapper's) heart—by design and upbringing—and eagerly participate in the age-old cycle of life and death each autumn.[394]

In "The Call of the Climb," Montanan Ruth Rudner says the ultimate quest for hunter and climber is the same. "Hunter and climber both enter an intimacy with the earth at the cost of often enormous physical and mental effort," she wrote. "It is an intimacy that challenges, demands, exhausts; that leads you on when you can go no farther; that presents a triumph as immense as it is private."[395]

I started hiking, climbing, and experiencing the wider wild world outside of hunting during the 1990s. In 1994, I climbed an active volcano, Mount Pacaya (8,350 ft.), in Guatemala; in 1996, it was Minnesota's highest point, Eagle Mountain (2,301 ft.) in the Boundary Waters Canoe Area Wilderness (BWCAW);[396] and in 1998, Mount Fuji (12,389 ft.) in Japan. Incidentally, the earliest recorded climb of a prominent mountain took place in A.D. 633 when a Japanese monk named En-no-Shokaku ascended Mount Fuji.[397]

According to the statisticians at CNN, our own Mount Hood (11,239 ft.) in Oregon is the second-most climbed mountain in the world after Mount Fuji.[398] In his book *Japan*, Richard Lloyd Parry describes Fuji's allure: "Even the rampant industrialization of post-war Japan hasn't been able to snuff out the magic of Mt. Fuji, which still must rate as one of the world's most beautiful mountains... No matter how many postcards you've seen, the sight of the pristine cone, dusted for most of the year with snow, and lightly scored with fissures and crevasses of ancient eruptions, is unfailingly moving and impressive."[399]

But Mount Fuji is not nearly as stunning up close (i.e., when climbing it) as it is from afar. The upper slopes are strewn with volcanic scree, which not only looks uninviting, but is difficult to hike-climb through. Still, without even recognizing it, Fuji and her brethren and mountain climbing were seeping into my soul, not unlike what must have happened to Reinhold Messner—the Lord of the Alps and Mount Everest—at a much younger age. Messner, the king of all climbers, says great things happen when man meets mountain.

Indeed, great things happened when Messner ventured into the high country, whether it was the Alps or the Himalayas, both of which he pretty much owned during the sixties, seventies, and eighties. In "Reinhold Don't Care What You

394 David A. Lien. "A Hunter's (& Trapper's) Heart." *Whitetales*: Winter 2008, p. 41.
395 David Petersen (ed.). *A Hunter's Heart*. New York: Henry Holt and Company, 1996, p. 49.
396 Minnesota's Eagle Mountain was my first state highpoint. I hiked it with a high school friend, Matt Washburn, on 5/22/96. Then, 11½ years later, I finished climbing the 50 state highpoints on 12,662-foot Borah Peak in Idaho. See article: "Lien climbs in all 50 states." Grand Rapids *Herald-Review*: 10/24/07, p. 3c.
397 Chris Case. "Museum to highlight evolution of mountain exploration." *Apex to Zenith*: Summer 2007, p. 28.
398 http://www.fourteenerworld.com/.
399 Richard Lloyd Parry. *Japan*. London: Cadogan Books, 1995, p. 277.

Think," *Outside* magazine contributor Brad Wetzler says, "No individual in history has done more to revolutionize climbing. Messner has surpassed even Sir Edmund Hillary in his genius for capturing people's imaginations and turning their eyes towards the snowy ramparts of the world's highest peaks."[400]

"Messner is to climbing what Michael Jordan is to basketball," Jon Krakauer (author of *Into Thin Air*) wrote in *Outside* magazine. "He has taken the sport to a level not previously imagined." German philosopher Arthur Schopenhauer once said, "Talent hits a target no one else can hit; Genius hits a target no one else can see." Edmund Hillary adds: "People do not decide to become extraordinary. They decide to accomplish extraordinary things," and Reinhold Messner was, much like Sir Edmund, an extraordinary mountaineering genius.

In "The Defiant One," *National Geographic Adventure* contributor John Rasmus explains that after Edmund Hillary, Messner is the world's most famous mountaineer. But among his climbing peers, Rasmus says, Messner stands alone. "He's the equivalent of Tiger Woods or Martina Navratilova: a brilliant, inventive athlete who revolutionized his sport through a combination of talent and monomaniacal pursuit of perfection—at any cost."[401]

In a sort of adventure junky's encyclopedia (LIFE Book's *The Greatest Adventures of All Time*), it's said that Messner's accomplishments rank as some of the greatest athletic endeavors in human history. In 1978, he and Austrian Peter Habeler became the first climbers to conquer Mount Everest without supplemental oxygen, something considered flatly suicidal at the time. Messner was also the first person to climb Mount Everest solo, which he did in 1980 on the North Face. Following his climb, there were no successful expeditions on the North Face for four years, though seven attempts were made.[402]

By bagging 27,824-foot Makalu and 27,923-foot Lohtse in a single Himalayan season in 1986, Messner became the first person to climb all 14 of the world's 8,000-meter (26,200-plus-foot) peaks. Considering that deaths among climbers of eight-thousanders is estimated at 3.4 percent annually, on his 29 expeditions to the world's 14 highest mountains, Messner ran a cumulative 98.6 percent risk of not coming back.[403] Indeed, as Reinhold documents in his book *All Fourteen 8,000ers*, in a mere two years—between 1982 and 1984—eight of the world's most experienced mountaineers died on big mountains.[404]

But in the overly modest words of Messner, "I didn't feel especially heroic to

---

400 Brad Wetzler. "Reinhold Don't Care What You Think." *Outside*: 10/4/02, p. 3.

401 John Rasmus. "The Defiant One." *National Geographic Adventure*: May 2004, p. 12.

402 Robert Sullivan and Robert Andreas (eds.). *The Greatest Adventures of All Time*. Des Moines, IA: LIFE Books, 2000, p. 107.

403 Reinhold Messner. *All Fourteen 8,000ers*. Seattle, WA: The Mountaineers, 1999, p. 186.

404 Reinhold Messner. *All Fourteen 8,000ers*. Seattle, WA: The Mountaineers, 1999, p. 220.

have climbed all fourteen of the eight-thousanders. Not exceptional in any way as a climber. I had seen something through, that was all, a task I had set myself four years before. I was pleased to [be]... done with it... It was all behind me at last. This was the morning I could start living the rest of my life. I felt light and free. The whole world lay before me."[405]

Messner's friend Gunter Sturm says for 16 years, Reinhold was a shaping force in Himalayan climbing: "Sixteen years of expedition climbing; that is sixteen years of being lonely, being constantly confronted by danger, anxiety and doubt; sixteen years of hardship and suffering."[406] Messner adds, "Personality alone is what counts, it's what gets you through it, along with the capacity for survival in increasingly more demanding situations."[407]

Since completing the eight-thousanders, Messner has taken himself to a new level, expanding his horizons beyond climbing and mountaineering. Among other adventurous endeavors, he spent the last years of the past century searching for Asia's fabled Yeti—a suitable mission for Messner (says Robert Sullivan in *The Greatest Adventures of All Time*): "One storied, near legendary mountain animal tracking another."[408]

Not to be outdone, Sir Edmund Hillary set out in search of the Abominable Snowman years before Reinhold, when he led a highly publicized 1960 Yeti expedition. "I am inclined to think that the realm of mythology is where the Yeti rightly belongs," Sir Ed said after weeks of hiking, climbing, and diligent investigation showed that alleged footprints and sightings all had mundane explanations, and relics of supposed Yeti skin and scalps actually came from bears and antelope.[409]

The Yeti's cousin, Sasquatch (or Bigfoot), is rumored to roam the wilds of North America, including some of Colorado's wildest public lands. When Julie Davis burst from her tent in a remote part of the San Juan National Forest late on the afternoon of August 5, 2000, she expected a confrontation with a hungry bear. What she got, says *Denver Post* contributor Theo Stein, was "the surprise of her life."[410]

Davis left Spring Creek Pass (10,901 ft.), which is traversed by State Highway 149, with a string of four pack goats and two border collies. She'd planned a 10-day trip down the Colorado Trail to Durango, and set up camp in a secluded

---

405 Reinhold Messner. *All Fourteen 8,000ers*. Seattle, WA: The Mountaineers, 1999, p. 11.
406 Reinhold Messner. *All Fourteen 8,000ers*. Seattle, WA: The Mountaineers, 1999, p. 15.
407 Reinhold Messner. *All Fourteen 8,000ers*. Seattle, WA: The Mountaineers, 1999, p. 222.
408 Robert Sullivan and Robert Andreas (eds.). *The Greatest Adventures of All Time*. Des Moines, IA: LIFE Books, 2000, p. 111.
409 Robert D. McFadden. "Sir Edmund Hillary, a Pioneering Conqueror of Everest, Dies at 88." *The New York Times*: 1/10/08.
410 Theo Stein. "Camper says she was 12 feet from curious giant." *The Denver Post*: 1/5/03.

meadow one night to tend a sick animal. After a day-plus of watching the increasingly nervous animals around her, it reached the point where, "I knew they were alarmed, and whatever it was it was right outside the tent."[411]

She yelled at her dogs to stay put, grabbed pepper spray, and rushed out to face the intruder. "In my wildest dreams, I might have expected to see a grizzly from the South San Juans," she said. "There is no question in my mind it was not a bear. I know from looking at the expression on its face and from the graceful way it ran off... You could see muscles moving under the fur... There's nothing more persuasive than staring something straight in the face."[412]

Colorado bow and elk hunting expert David Petersen, who has logged more time in the San Juan Mountains than any person I know, had a similar experience one late September evening, as recounted in his book *On The Wild Edge*: "I approached a right-hand bend in the trail and was startled to see, coming around the bend toward me, just twenty yards below, the hulking form of some two-legged shadowy thing. In the same instant, whatever it was saw me. We both stopped short and stared, though I could see no glint of eye in the hairy face..."[413]

"The idea of someone headed uphill and into the wilds on a chilly, moonless night, without a flashlight or gear of any kind, seemed nonsensical. And if the being was human, why was he acting so... strange? My next thought was *bear*... how often do bears stroll around on their hind legs in the dark? I almost called out, 'Hello?' But for reasons I can't explain, I held off... Not wanting to startle the creature, I moved slowly to extract an arrow from the quiver and place it on the bow string. But even that was enough to prompt Bigbutt, obviously the nervous sort, to shuffle to the edge of the old road cut—a sudden move that made me jump and set my heart to thumping.[414]

"With an arrow strung and the creature now standing at the edge of the woods, swaying slowly back and forth as if trying to get a better sense of me, I reached for my flashlight. This second movement, small as it was, was enough to prompt Bigbutt to lurch into the blackness of the adjacent trees and brush... Incredibly, this beast the size of a bear or a man had made no sound moving into or through the dry woods... I did see *something* out there—something large and very alive, something sentient and willful and beyond logical comprehension. While I have no doubt that a perfectly pragmatic explanation exists, I doubt I'll ever find it."[415]

411 Theo Stein. "Camper says she was 12 feet from curious giant." *The Denver Post*: 1/5/03.
412 Theo Stein. "Camper says she was 12 feet from curious giant." *The Denver Post*: 1/5/03.
413 David Petersen. *On The Wild Edge*. New York: Henry Holt and Company, 2005, p. 78.
414 David Petersen. *On The Wild Edge*. New York: Henry Holt and Company, 2005, p. 79.
415 David Petersen. *On The Wild Edge*. New York: Henry Holt and Company, 2005, p. 79.

Reinhold Messner, the Sasquatch of mountaineers—an elusive loner with near-mythic abilities—is a difficult beast to follow too. Since 1986, only a handful of climbers have repeated Messner's feat of summiting the world's 14 highest mountains, and only one American (Ed Viesturs) as of the dawn of the twenty-first century. I can't say I'll ever be well-known for mountaineering or anything else I do, but I too believe great things happen when "man meets mountain." Indeed, whenever people venture into any wild place, they come away with a sense of inner peace and wonder completely lacking in our now mostly urban and supposedly civilized ("civilization-ravaged," I say) society and lives.

> "He who feels the spell of the wild, the rhythmic melody of
> falling water, the echoes among the crags, the bird-songs,
> the wind in the pines and the endless beat of waves upon
> the shore, is in tune with the universe."
> —Enos Mills[416]

MAY

Hiking and climbing in the Rocky Mountains, or any other mountains, is not something I ever dreamed about or aspired to do as a boy or young man. Like drinking beer, I suppose, it was an acquired taste, and after the first round, a few more seemed like a good idea. So, after hiking-climbing eight 14ers between August and November 1999, I impatiently waited for winter to wane and spring to arrive. In late May, the start of my 2000 climbing season took the form of Mount Antero in the Sawatch Range.

Due to the lingering mountain snowpack, May can be a bit early for nontechnical 14er climbing, but I was anxious to get started again after a long winter layover. Like Mitch Friedman (executive director of Conservation Northwest) said, "Summer approaches, and the restlessness grows in my legs to feel the burning sensation of the uphill slog. The longing grows in my lungs to fill with pure mountain air. My eyes crave expansive vistas. My nerves yearn to tingle with the prospect of a bear roaming in the brush."[417]

I was the only person experiencing Antero's pure mountain air on this day. Not even another track (man or bear) in the snow! The narrow national forest road leading to Antero was still snow-covered in places, but I managed to reach a suitable car-camping location just before it became impassable. Mount Antero attracts the attention of hikers, climbers, and others for many reasons,

416 Enos Mills was a Colorado naturalist, explorer, and photographer who was instrumental in persuading Congress to create Rocky Mountain National Park, in 1915.
417 Mitch Friedman. "Celebrating Our Remaining Truly Wild Country." *Conservation Northwest*: Summer 2008, p. 3.

according to Summitpost.org.

First off, it's the 10th highest mountain in Colorado, so peak baggers can take in some lofty views. A wealth of fine gemstones have also been plucked and mined from its slopes. This makes Antero popular with rock hounds. Last, for motorized wreckreationists, there's a good 4WD road (an oxymoron) on the peak's west side that ends not far from the summit, at 13,700 feet, but using the quads God gave us is always the best mode of transport for experiencing Antero or any other mountain.[418]

The 4WD road runs from the Baldwin Gulch trailhead to within a few hundred vertical feet of the summit, covering a distance of nine miles. It was built in the early 1950s by a company looking to mine beryl on Antero.[419] The beryl mining proved uneconomical, but the road is still being used. Depending on your perspective, this is a great convenience or a blight on an otherwise beautiful mountain.[420] Considering that today (as previously mentioned), only 8 percent of the national forest acreage in Colorado lies beyond one mile of a road, and only 4 percent for BLM lands, "blight" is right.

Colorado's long history of mining has regrettably left many such alpine landscapes scarred by roads. Mount Antero's mining history kicked off in 1881 when a Salida man, Nelson Wanemaker, discovered gems high on the mountain. His discovery was publicized a few years later and Mount Antero quickly became a famous collecting destination for aquamarine, phenacite, fluorite, topaz, and smoky quartz crystals. With most discoveries occurring above timberline, this is the highest known gem locale in the United States.[421]

According to the Denver Museum of Nature & Science, 35 million years ago, mineral-rich magma under high pressure intruded from the fiery depths into rock layers of the Sawatch Range. As the magma cooled, it trapped cavities of mineral-saturated fluid heated to temperatures as high as 600° C (1,112° F). Minerals crystallized out of that fluid, and over millions of years, the mountains were uplifted and eroded, exposing the gems within.[422]

Mount Antero is also known for being the highest Colorado mountain named after a Native American. Neighboring peaks Tabeguache, Shavano, and Ouray have Indian names too. Chief Antero was the leader of the Uintah band of Ute Indians and a proponent for peace between the Utes and white men during an uprising that occurred in the late 1860s and early '70s. He was also one of the

418 http://www.summitpost.org/show/mountain_link.pl/mountain_id/204.

419 Beryl is the principal ore of the metal known as beryllium. Beryllium is lighter than aluminum and stronger than steel. Today, beryllium alloys are used in atomic reactors, electrical components, and on spacecraft.

420 http://www.summitpost.org/show/mountain_link.pl/mountain_id/204.

421 http://www.summitpost.org/show/mountain_link.pl/mountain_id/204.

422 Denver Museum of Nature & Science. "Largest Aquamarine Ever Found in North America Donated to Museum." www.dmns.org.

signers of the 1880 Washington Treaty, which led to the Utes losing most of their land.[423]

In her article "Naming the Indian group of the Sawatch Range," *Colorado Central Magazine* contributor Virginia Simmons says the Sawatch Range name derives from a Ute word, *saguguachipa*, meaning "blue earth." In the Ute's cosmology, the mountains and valleys where they lived were part of the Blue Earth, or Middle Earth, in contrast to the Lower Earth of deep canyons and the Upper Earth of mountaintops. Among the Ute's favorite places in Middle Earth was Colorado's Upper Arkansas River Valley, where they frequently set up camp.[424]

Virginia explains that the first recorded use of the phrase "Sahwatch Range" came from a report of Captain John Gunnison's expedition in the 1850s. It was written by Lieutenant E. G. Beckwith, who was actually referring to the mountains west of the San Luis Valley (i.e., the San Juans). But the name was later applied to the more northerly chain with its glorious fourteeners. The Sawatch Range's most southerly peaks (which have been popularly called the Indian Group) are Mount Shavano, Tabeguache Peak, and Mount Antero.[425]

After spending the night car-camping amidst the Ute's forfeited lands on Antero's still mostly snow-covered lower slopes, I was up and following the path of least resistance (the one nearest to camp) before dawn, but soon surmised that it wasn't leading to the summit. Due to the snow cover and my unfamiliarity with the area, following the standard Antero route became a guessing game, and I guessed wrong.

Instead of backtracking and wasting valuable time and elevation, I decided to leave the trail behind and bushwhack up the mountainside, not following a trail but making my own—the best way to go through life sometimes. As Muriel Strode said, "Do not follow where the path may lead. Go instead where there is no path and leave a trail." Robert Frost adds, "And I took the road less traveled, and that made all the difference."

Even Jesus seems to agree with Muriel's and Robert's sentiment: "Go in through the narrow gate; because broad and spacious is the road leading off into destruction, and many are the ones going in through it; whereas narrow is the gate and cramped the road leading off into life, and few are the ones finding it" (Matthew 7:13-14; John 14:6). Sound advice, Jesus.

After reaching the summit of a 13,762-foot peak north of Mount Antero, I stopped to don heavy gloves, goggles, and a facemask before continuing

---

423 http://www.summitpost.org/show/mountain_link.pl/mountain_id/204.

424 Virginia McConnell Simmons. "Naming the Indian group of the Sawatch Range." *Colorado Central Magazine:* June 2005.

425 Virginia McConnell Simmons. "Naming the Indian group of the Sawatch Range." *Colorado Central Magazine:* June 2005.

up Antero's wind-blasted, snow-drifted north ridge through streaming clouds, a strange and exhilarating experience. Upon summiting, I was greeted by partly cloudy but clearing skies. A combination of bright sun, blue sky, and backlit clouds made for some of my favorite 14er summit photos, and not another soul in sight! It was a glorious, maybe even heavenlike, morning.

Montanan Ruth Rudner knows the feeling: "On the mountain the sun shifts. The shadows slant longer across the face. A stone falls, arcing out, away from the mountain, to land on eons of fallen stones below. It is so easy to let go, to fall... A tiny flower grows in a crevice just above your foot. The rock eases into dirt. A tuft of goat hair clings to a scraggly twig. A cloud hovers over the sun, then passes by. You reach the top. The mountain holds you. You have come to that place where you were going."[426]

Tibetan monks talk about a state of mind called "bardo," where you are literally between two existences. Traveling solo in the wilds (explains *Boundary Waters Journal* contributor Dick Pula) helps us find that balance between the modern and real worlds. The monks also talk about "chod," or facing fears and achieving a trust of life that finally means making peace with reality.[427] It's on fourteeners and in other wild places where, I think, bardo and chod are best experienced.

There is something almost beyond words about being alone in such places that makes the experience especially meaningful. So much that is deeply felt is, in its very nature, undefinable, John Muir said. Especially when we travel alone, I say. While climbing through the scraggily trees, loose scree, shifting boulders, and snow and clouds on this wild and wind-ripped mountainside, I was in a world where the things that mattered most were the pack on my back, the sunlight above, and the look of the way ahead.

It was a world where I relied only upon myself, one that allowed me to escape, albeit temporarily, from the persistent problems of everyday life; work problems, personal problems, friends' problems, coworkers' problems. While up there, closer to heaven (and further from hell) than most will ever get, all problems beyond the immediate difficulties of hiking, climbing, and survival fade into the background, becoming part of life's irrelevant clang and clutter.

One of the keys to keeping sanity intact, I've found, is focusing on potential adventures and opportunities, while not getting bogged down in the daily 9-to-5 grind and its always immediate problems. Problems scream at us, but opportunities only whisper. Problems are usually deemed urgent, while opportunities are merely important. If we let problems dominate our lives, opportunities to

---

426 David Petersen (ed.). *A Hunter's Heart*. New York: Henry Holt and Company, 1996, p. 50.
427 Dick Pula. "Solo Ice Travel." *The Boundary Waters Journal*: Winter 2009, p. 70.

experience wild places and expand our horizons begin to fade away. Turn that around and make opportunities the focus, dealing with the problems but not letting them take over.[428]

Like Robert Service said, "Be master of your petty annoyances and conserve your energies for the big, worthwhile things. It isn't the mountain ahead that wears you out—it's the grain of sand in your shoe." Management guru Stephen R. Covey adds, "Let's climb out of the crevasses of the 'day to day' so we can see things from a mountaintop perspective." After getting such a perspective from the summit of Mount Antero (my 9th fourteener), on May 20, I took a break from climbing 14ers due to my cousin Karl Nyman flying out from Minnesota for a Rocky Mountain Memorial Day weekend car-camping excursion.

Another opportunity to leave problems behind and focus on what's most important and worthwhile in life; experiencing and exploring (and then endeavoring to defend) the wild and natural world. I had a whirlwind tour of two mountain ranges (the Front Range and the Sangre de Cristos) encompassing some of the state's finest 14er-studded terrain planned for Karl's short visit.

On our way to the Great Sand Dunes National Monument in the Sangre de Cristos (not far from New Mexico in south central Colorado), we passed the easternmost point of the Rocky Mountains, the Spanish Peaks. In "Magical Mountains," *Pueblo Chieftain* contributor Scott Smith says these peaks stretch skyward like a pair of regal, rocky sentinels, marking the place where the plains end and the mountains begin. They're a matched set, an alpine version of yin and yang, also known by their American Indian name, *Wahatoya*, which translates to "breasts of the Earth."[429]

Smith explains that these towering double Ds—visible from as far away as Colorado Springs (about 100 miles to the north) on some days—have been among the Southwest's most prominent landmarks for a long time. They served as natural signposts that guided American Indians, Spanish and French explorers, and finally a flood of wagon train settlers. A glimpse of the Spanish Peaks was often the first confirmation for wagon trains venturing west that they were about to get up close and personal with the highest, wildest part of the great unknown that lay ahead: the Rocky Mountains.[430]

West Spanish Peak (13,626 ft.) is 374 feet shy of ranking as one of Colorado's fourteeners. Consequently, it doesn't get the same use and abuse as its higher profile brethren, but that doesn't mean climbing her isn't well worth the time and effort. It is, but that would have to wait for another day and out-of-town visitor, <u>Todd Laughman</u>, during 2005. Karl and I passed by the Spanish Peaks and over

428 Rick Olson. "Leaders need to stay motivated, too." *Credit Union* magazine: October 2002, p. 24.
429 Scott Smith. "Magical Mountains." *The Pueblo Chieftain*: 10/6/05.
430 Scott Smith. "Magical Mountains." *The Pueblo Chieftain*: 10/6/05.

the Sangre de Cristos on U.S. Highway 160's La Veta Pass (9,314 ft.), leaving the Great Plains behind and entering into the Rocky Mountain's San Luis Valley.

Highway 160 continues west across the valley into the San Juan Mountains, connecting Walsenburg and eastern Colorado with Alamosa, Monte Vista, and South Fork. West of South Fork it crosses over Wolf Creek Pass (10,863 ft.), the highest mountain pass on the Continental Divide, on its way to Pagosa Springs and Durango, then on into New Mexico and Arizona, but we were headed north to Colorado's Great Sand Dunes.

On the eastern side of the relatively remote and sparsely populated high-desert San Luis Valley (the largest alpine valley above 7,500 feet in the world), sitting between the 14er-studded Blanca Massif and Crestone Peak and Needle, you encounter the Great Sand Dunes, the tallest in North America: 39 square miles of sand rising to a height of almost 750 feet above the valley floor.[431] The Dunes are a product of wind and rain eroding the San Juan Mountains and Sangre de Cristo Mountains, which straddle the San Luis Valley on the west and east respectively.

The Great Sand Dunes formed here because prevailing winds blow across the valley floor meeting winds coming from the opposite direction. The sand drops to the valley floor and is held in place by water flowing from the Sangres, much of it in Medano Creek, which exhibits a unique water-and-sand-based phenomenon called *surge flow*.[432] These surges may remind you of waves at a beach, says a Saguache County Visitor's Guide. Each time a surge occurs, a mound of underwater sand (which has temporarily dammed some of the water in the creek bed) collapses. Surge flow occurs in only a few places on earth, and none are as accessible as the Great Sand Dunes.[433]

After visiting the Dunes (which were designated a national park in 2004), we headed north into the Front Range and Rocky Mountain National Park northwest of Boulder, spent an afternoon and night in the park, then drove south to 14er Mount Evans. The Mount Evans Highway starts about 35 miles west of Denver and twists its way south from Idaho Springs through the Arapaho National Forest past alpine lakes, bristlecone pines, and occasional bighorn sheep and mountain goats to just below the summit of Mount Evans, a mountain with one of the best, most easily accessible fourteener summit views in Colorado.[434]

Most of the Mount Evans massif is now part of the 74,401-acre Mount Evans Wilderness in the Arapaho and Pike National Forests. And as told by *Desert Morning News* contributor Lynn Arave (in "The high road"), the Mount Evans Scenic

431 http://www.sangres.com/nationalparks/sanddunes/.
432 http://www.sangres.com/nationalparks/sanddunes/.
433 Saguache County 2008 Visitor's Guide (p.7): www.saguachecounty.net.
434 David Olinger and Steve Lipsher. "Scenic lands tagged for sale." *Denver Post*: 2/19/06, p. 1A.

Byway (Colorado Highways 103 & 5) is the highest paved passenger route in North America. It ends at 14,130 feet above sea level, or 20 feet higher than the better-known Pikes Peak Highway, which is still (but not for much longer) a gravel road on its upper reaches. The Mount Evans road is narrow, but well-suited for passenger cars, and climbs through more than 7,000 feet on its 28-mile run.[435]

The road also facilitates easy access to the tippy-top of Mount Evans (14,264 feet), Colorado's 14th highest peak, located about 30 miles as the crow flies southwest of Denver. The Mount Evans road ends at a small parking lot, where it's a quarter-mile hike through another 134 feet to the rocky summit. "At the summit you'll enjoy the big picture–the entire Front Range sprawls at your feet," exclaims the official website of Colorado's scenic and historic byways.[436]

According to the website, "This highest of Rocky Mountain Highs brings you to the rarefied world above timberline, a singular amalgam of hardy wildflowers, lichens and grasses, furry mammals, like pikas and marmots, rock-jumping mountain goats and alpine lakes. This is perhaps the best place in Colorado to catch a glimpse of the stately bighorn sheep."[437] Even downtown Denver can be spotted on clear days, which isn't much to look at given the metro area's frequent puke-yellow smog layer.

An average of 473 vehicles a day used the Mount Evans Highway during the summer of 2006, up from a 418 vehicle average during 2005, and their tailpipe pollution alone may obscure your view of the astounding surrounds. A $10 vehicle entrance fee is charged to travel the upper 14 miles of the Mount Evans Byway above its Echo Lake junction, but there's another, nerve-racking price to pay for some–the steep, narrow road (though fully paved) has no guardrails.[438]

Lynn Arave says this may tax even the only mildly acrophobic drivers and passengers, and some people have been known to park and walk the final four miles. Whether walking or driving, about 1½ road miles below Mount Evan's summit you'll pass a spot that has special meaning to Utahns, like Lynn. The road's elevation here is 13,528 feet, the height of Utah's highest point–Kings Peak–which is accessible only by hiking-climbing 28½ miles (round-trip) through 5,350 feet of elevation gain in the Uinta Mountains.[439] It's a highpoint that I

435 Lynn Arave. "The high road: Driving to the top of Colorado's Mount Evans takes your breath away." *Desert Morning News*: 8/13/06.

436 Lynn Arave. "The high road: Driving to the top of Colorado's Mount Evans takes your breath away." *Desert Morning News*: 8/13/06.

437 Lynn Arave. "The high road: Driving to the top of Colorado's Mount Evans takes your breath away." *Desert Morning News*: 8/13/06.

438 Lynn Arave. "The high road: Driving to the top of Colorado's Mount Evans takes your breath away." *Desert Morning News*: 8/13/06.

439 Lynn Arave. "The high road: Driving to the top of Colorado's Mount Evans takes your breath away." *Desert Morning News*: 8/13/06.

completed in a marathon-like (for me) 14½ hours during October 2005.

With a dozen summits measuring more than 13,000 feet high, the Uintas are Utah's tallest mountain range. The mountains, a middle range of the Rocky Mountains, run south of the Wyoming border, covering about 150 miles between Kamas and Flaming Gorge, Utah, making this one of North America's few large east-west ranges. The 460,000-acre High Uintas Wilderness Area was designated by Congress in 1984.[440]

Construction on Colorado's Mount Evans Highway began in 1922, was completed in 1930, and the road was first paved a few years later. Mount Evans is named after John Evans, Colorado's second territorial governor. The peak was originally named Mount Rosalie, after the wife of painter Albert Bierstadt, but in 1895, Colorado's legislature officially changed its name.[441] Temperatures of 50 degrees below Fahrenheit and wind speeds approaching 200 mph have been recorded on Mount Evan's summit, so bring a jacket.[442]

Fourteenerworld.com says Mount Evans has seen cosmic ray, space, and aviation experiments, testing of the effects of nuclear explosions on the atmosphere, and meteorological observations and surveys. Many famous scientists from all over the world have studied/worked on Mount Evans, including Carl Anderson, R. A. Millikan, and A. H. Compton, three of the nation's Nobel Prize winners in physics.[443]

Summitpost.org recommends the Red Ram restaurant and saloon in Georgetown for some cheap eats and a few brews after driving or climbing Mount Evans.[444] Karl and I skipped the Red Ram and returned to Colorado Springs, where Karl prepared to head for home in Minnesota while I set my sights on the Sawatch Range's roadless and mostly tailpipe pollution-free Tabeguache Peak.

JUNE

Mount Shavano (14,299 ft.) dominates the skyline 15 miles northwest of Salida, Colorado. Tabeguache Peak (14,155 ft.)—the 26th highest mountain in Colorado—is less than a mile away from Shavano and wasn't considered a separate peak until 1931.[445] Tabeguache and Shavano are the southernmost fourteeners in the Sawatch Range, and they're normally climbed together, being separated by a mile-long ridge with a 250-foot drop in between. I chose to tackle them sepa-

440 Rachel Carley. *Wilderness A to Z.* New York: Simon & Schuster, 2001, p. 300.
441 http://www.fourteenerworld.com/.
442 Lynn Arave. "The high road: Driving to the top of Colorado's Mount Evans takes your breath away." *Desert Morning News:* 8/13/06.
443 http://www.fourteenerworld.com/.
444 http://www.summitpost.org/mountain/rock/150481/mount-evans.html.
445 Ed Quillen. "A mountain by any other name would soar as high." *Colorado Central Magazine:* May 1998.

rately, skipping the surely spectacular but taxing ridgeline traverse.[446]

Colorado Central Magazine contributor Virginia Simmons explains (in "Naming the Indian group of the Sawatch Range") that the name "Tabeguache" first appeared in print in 1925 in the Colorado Mountain Club's periodical *Trail or Timberline*, according to John L. Jerome Hart's book *Fourteen Thousand Feet: A History of the Naming and Early Ascents of the High Colorado Peaks*.[447]

Although Tabeguache has had many spellings over the years, based on phonetics the word seems to have been derived from a Ute Indian term, *Mogwatavungwantsingwu*, which means roughly "cedar-bark, sunny slope people." Simmons says the Ute tribe had several bands in Colorado, Utah, and New Mexico, but the Tabeguache Band was the largest, consisting of 500 to 1,000 Native Americans who came and went in extended family groups.[448]

The Ute people, who are the oldest continuous residents of Colorado (notes the aforementioned Saguache County Visitor's Guide), probably arrived in the San Luis Valley area around 1,300 AD. Imagine how their culture changed when the Spanish brought horses to the New World. The expansion of their hunting range and ability to more easily move nomadic camps made life much easier. When European settlers first arrived in the Saguache County region, the Utes were making annual trips over the passes of the Sangre de Cristos to hunt buffalo on the eastern plains.[449]

Being hunter-gatherers, Tabeguaches roamed a territory that includes the Gunnison Valley, Uncompahgre Valley, San Luis Valley, the Upper Arkansas River Valley, and South Park. The Upper Arkansas is where they often hunted, gathered food, and fought when encountering Plains tribes or eastern militias. The natural resources of this region made it an especially popular place, with ample grass for horses, plentiful timber, pinyon nuts, berries, seeds, game, hot springs, a relatively temperate climate, and extraordinary scenery. When pioneers first arrived, the Utes they met were usually Tabeguaches.[450]

After several hours of struggling up Mount Tabeguache's steep southwest ridge on the Jennings Creek Trail, a route with no switchbacks or other trail improvements, I reached the summit exhausted. This is a terrible trail with serious erosion problems, and the Forest Service wisely chose to close it dur-

446 Randy Jacobs (ed.). *Guide to the Colorado Mountains*. Golden, Colorado: The Colorado Mountain Club Press, 2000, p. 197.

447 Virginia McConnell Simmons. "Naming the Indian group of the Sawatch Range." *Colorado Central Magazine*: June 2005.

448 Virginia McConnell Simmons. "Naming the Indian group of the Sawatch Range." *Colorado Central Magazine*: June 2005.

449 Saguache County 2008 Visitor's Guide (p.12): www.saguachecounty.net.

450 Virginia McConnell Simmons. "Naming the Indian group of the Sawatch Range." *Colorado Central Magazine*: June 2005.

ing September 2002. To climb Tabeguache on the standard route today, first you summit Shavano on the Blank Gulch Trail, then ridgeline traverse over to Tabeguache, retracing your steps on the return trip.

From the summit of Tabeguache, nearby Mount Shavano beckoned, but I was spent and decided a *twofer* (bagging two 14ers in a day) wasn't an option. During the descent, I spied a mountain goat retreating down the mountainside, my first wild mountain goat sighting, and noticed dozens of bees taking advantage of the wildflowers blooming on the sunny but exposed windswept ridges.

I thought it strange, at the time, to see bees buzzing around so high in the mountains. Gale-force gusts could easily put an end to their days in such an open-to-the-elements location, but like frogs croaking away through warm summer nights, unwittingly calling predators near, they do what they do. Is it possible that bees occasionally cruise the high ridges for pure pleasure and enjoyment, maybe even the exhilaration or danger of it? Could their species also have wanderers and adventurers? Maybe I was in the company of the Reinhold Messner of bees. I think not, but it doesn't hurt to ponder and hope.

In all likelihood, my bee companions were there due to simple evolutionary programming, but Edward Abbey has another theory, as this excerpt from *Desert Solitaire* suggests: "See the frogs clinging to the edge of their impermanent pond... all croaking away in tricky counterpoint... Why do they sing? What do they have to sing about?... It may nevertheless be the case that these small beings are singing... out of spontaneous love and joy... Where there is no joy there can be no courage; and without courage all other virtues are useless."[451]

After climbing Tabeguache, I returned to the Sangre de Cristos and Blanca Peak, not far from the Great Sand Dunes National Monument. Summitpost. org notes that Blanca Peak (14,345 feet high and the 4th tallest mountain in Colorado) is part of the impressive Rocky Mountain battlement that makes up the Sierra Blanca Massif. Blanca is the highest point of this massif, which includes Little Bear Peak (14,037 ft.), California Peak (13,849 ft.), Mount Lindsey (14,042 ft.), the Iron Nipple (13,480 ft.), Hamilton Peak (13,658 ft.), Peak 13,828, and Ellingwood Point (14,042 ft.).[452]

Blanca Peak is also a county "highpointers'" dream: it's the highpoint of Alamosa, Costilla, and Huerfano counties, the only such tricounty highpoint in the nation. In addition, Blanca is the highest peak in Colorado with a "technical" route on it: Blanca's northeast face is a mile wide and one of Colorado's largest walls.[453] The north face of Blanca Peak also shades the southernmost glacier in the United States, which—like all Colorado glaciers, I'm guessing—is so

451 Edward Abbey. *Desert Solitaire*. New York: Simon & Schuster, 1968, p. 125.
452 http://www.summitpost.org/show/mountain_link.pl/mountain_id/312.
453 http://www.fourteenerworld.com/.

small (and shrinking) that it's not even recognizable to your "average Joe" peak bagger like me.[454]

Speaking of highpoints, each state has a place of highest natural elevation (explains *New York Times* contributor Jane Margolies, in "All Peaks, No Valleys"), ranging from the piddling 345-foot Britton Hill in Florida to 20,320-foot Mount McKinley in Alaska. Some sites are known as "flip-flop" highpoints, Margolies says, because visitors can drive up in a car and hop out in sandals to pose by a marker. Others require multiday climbs, but all are important to the increasing number of "highpointers."[455]

An estimated 10,000 people are caught up in the hobby, which blends (Jane says) the rigors of adventure travel with the fastidiousness of stamp collecting. "Highpointers hopscotch the country, then go back home where they track their accomplishments on spreadsheets and wall maps bristling with color-coded pushpins, and then plot their next outings."[456] There aren't any wall maps or color-coded pushpins in my house, but spreadsheets do come in handy for keeping track of completed highpoints and other peaks, as does the Highpointers Club.

The club is a volunteer-run organization with a propensity for substituting "k" for "c" in its literature—as in "Keep Klimbin'," the club's motto—in honor of its founder, Jack Longacre, who died in 2002 and whose ashes have been scattered on all the state highpoints.[457] Most devoted highpointers also internalize the "three Ks" to climbing: Klimb high, Klimb often, Klimb safely! While many sedentary, less adventurous types may wonder about the real or perceived need to "track down some featureless bump in the middle of nowhere," highpointers will tell you the hobby provides much more than just the satisfaction of coloring in all the pixels on a map back home.[458]

It gets us off the well-trodden paths and into the nooks and crannies of America, and Blanca Peak is one such worthy nook and cranny, which I'll get back to in a minute.[459] Kevin Baker is a Colorado-based highpointer, "someone whose mountaineering résumé details his quest to conquer the highest elevation point in a given area," notes CMC member Lori Spaulding. Such places vary from U.S. state highpoints to Colorado's county highpoints to national park high points, and more.[460]

Kevin's highpointing accomplishments include more than 600 U.S.-ranked (higher than 299 feet of prominence) summits, all of Colorado's 14ers and

454 Richard D. Lamm. *Mountains of Colorado.* Chambersburg, PA: Graphic Arts Center Publishing Company, 1999.
455 Jane Margolies. "All Peaks, No Valleys." *The New York Times*: 11/16/08.
456 Jane Margolies. "All Peaks, No Valleys." *The New York Times*: 11/16/08.
457 Jane Margolies. "All Peaks, No Valleys." *The New York Times*: 11/16/08.
458 Jane Margolies. "All Peaks, No Valleys." *The New York Times*: 11/16/08.
459 Jane Margolies. "All Peaks, No Valleys." *The New York Times*: 11/16/08.
460 Lori Spaulding. "Mountaineer on a quest to conquer highest points." *PikesPique*: May 2010, p. 1.

county highpoints, 171 county highpoints nationwide, 91 of the 100 highest highpoints, all ranked summits in El Paso and Teller counties, and 196 of Colorado's ranked 13ers. "Highpointing gets you out in different areas you wouldn't go to normally," says Kevin, who also serves on the board of directors of the State High Points Club based in Golden, Colorado.[461]

Kevin considers 13,804-foot Gannet Peak among the hardest state highpoints, and I'll second Kevin. Tucked into Wyoming's Wind River Mountains, Gannett is "the most remote of all the state highpoints and a 40-mile roundtrip backpack involving glacier travel, stream crossings, mosquito tolerance, scrambling, a couloirs climb and an exciting finish along the exposed summit ridge."[462] After returning from climbing Gannett in 2007, I e-mailed this postclimb trip report to friends and family:

> "We reached Gannett Peak on Tuesday, 24 July, after hiking and climbing some 40-plus miles (round-trip) through 8,650 feet of elevation gain over six days (from the Elkhart Park trailhead) hauling 50-60 lb packs. We also had a fairly brutal, but beautiful 14-hour summit day that included roped glacier travel, ice climbing, and rappelling.
>
> "Three of our five team members (including me) made the summit. Two turned back en route. I never imagined there were still glaciers like we saw from Gannett's summit in the Lower 48. We camped near Island Lake our first night out, then on Bonney Pass (12,800 feet) the second and third nights, and near Freemont Crossing during our last night."

According to the Highpointers Club, the earliest reference found related to state highpoints was a 1909 article in *The National Geographic Magazine* titled "The Highest Point in Each State." At the time, locations of some state highpoints were not known, the elevations of several others were listed as approximate, and some highpoints listed were later discovered to be incorrect.[463] Over 100 years later, in 2010, 12-year-old Boulder alpinist Matt Moniz and his father, Mike Moniz, set out to climb the highest point in each state in 50 days.[464]

The pair stood atop their 50th highpoint, Hawaii's Mauna Kea, on July 16, 2010, stopping the clock at 43 days, 3 hours, 51 minutes, and 9 seconds, a new speed record for highpoints—which are generally self-reported to the

---

461 Lori Spaulding. "Mountaineer on a quest to conquer highest points." *PikesPique*: May 2010, p. 1.

462 Lori Spaulding. "Mountaineer on a quest to conquer highest points." *PikesPique*: May 2010, p. 1.

463 Highpointers Club. "HP Club Alert." *E-mail*: 4/16/09.

464 Jenn Fields. "Boulder 12-year-old Matt Moniz, dad break speed record for states' high points." *ColoradoDaily.com*: 7/21/10.

Highpointers Club but not verified by them.[465] It took me a much more leisurely 11½ years to complete them, starting with Minnesota's Eagle Mountain (2,301 ft.) on 5/22/96, and ending with Idaho's Borah Peak (12,662 ft.) on 9/22/07.[466] Back to Colorado and Blanca Peak.

The first recorded Blanca Peak ascent was on August 14, 1874, by the Wheeler Survey, but to their surprise, they found evidence of a circular excavation on the summit, possibly built by wandering Spaniards, but more likely by Ute or Navajo Indians. Blanca Peak is known to the Navajo as *Tsisnassjini*, or White Shell Mountain, and it's their Sacred Mountain of the East. The Navajo have four such sacred mountains: two are in Colorado, one in Arizona, and the other in New Mexico.[467]

The Navajo believe their Creator placed them on the land between the four mountains, representing the four cardinal directions: Blanca Peak is the Sacred Mountain of the East; Mount Taylor (*Tsoodzil*, Blue Bead or Turquoise Mountain) is the Sacred Mountain of the South, located north of Laguna, New Mexico; San Francisco Peak (*Doko'oosliid*, Abalone Shell Mountain) is the Sacred Mountain of the West, near Flagstaff, Arizona; Mount Hesperus (*Dibé Nitsaa*, Big Mountain Sheep), also called the Obsidian Mountain, is the Sacred Mountain of the North, found in Colorado's La Plata Mountains (an arm of the mighty San Juans).[468]

During 2004, I climbed one-third of the Sacred Mountain of the West, Humphreys Peak, in Arizona. The San Francisco Peaks consist of three main summits: Humphreys Peak (12,633 feet), Agassiz Peak (12,300 feet), and Fremont Peak (11,940 feet). They're a volcanic range located in north central Arizona near Flagstaff. The highest summit in the range, Humphreys Peak, is also the highest peak in Arizona. The San Francisco Peaks (known locally as "the Peaks") are the remains of an eroded stratovolcano.[469]

On to Colorado's Sacred Mountain of the East. The Lake Como road is the standard approach route for Ellingwood Point, Blanca Peak, and Little Bear Peak, and this 4-wheel-drive "road" is purportedly Colorado's toughest. Having inched and scraped my way up it as far as a Blazer would take me, leaving more than a little metal on rocks and ledges along the way, I know from experience that hiking (not driving) the lower sections of this road is recommended. But the payoff was car-camping on an airy switchback with an unimpeded sunset view of the San Luis Valley.

465 Jenn Fields. "Boulder 12-year-old Matt Moniz, dad break speed record for states' high points." *ColoradoDaily*.com: 7/21/10.

466 See articles: "Lien climbs in all 50 states." Grand Rapids *Herald-Review*: 10/24/07, p. 3c & "Milepost #50 for David A. Lien." *Apex to Zenith*: Winter 2007, p. 4.

467 http://www.summitpost.org/show/mountain_link.pl/mountain_id/312.

468 http://www.lapahie.com/Sacred_Mts.cfm.

469 http://en.wikipedia.org/wiki/San_Francisco_Peaks.

John Fiedler and Mark Pearson have described the San Luis Valley about as good as anyone in their book *Colorado's Wilderness Areas*. The valley, they explain, lies in the rain shadow of the San Juan Mountains and, consequently, receives only about eight inches of precipitation annually, making it one of Colorado's driest deserts. In midsummer, heat from the valley's sun-baked sands can quickly penetrate even the thickest-soled shoes. In the winter, valley residents experience some of the state's coldest temperatures, with lows frequently dipping to minus 30 and 40 degrees Fahrenheit.[470]

High above the San Luis Valley, I was up and climbing before first light and above tree line by sunrise, the best place to be for that always momentous event in the mountains. As dawn crept slowly across the still-slumbering valley below, I climbed towards Blanca Peak, taking notice of two climbers hundreds of feet above me traversing a ridge that would take them to Ellingwood Point. I wouldn't climb Ellingwood until 2001, but Blanca would feel the weight of my bootsteps today.

Blanca Peak looks particularly imposing when viewed from below and at a distance, conjuring up images like those described by Reverend S. Hall Young (a friend of John Muir). Young said, "The earth-fires that had melted and heaved it, the ice mass that chiseled and shaped it, the wind and rain that corroded and crumbled it, had left plenty of bricks out of that battlement, had covered its face with knobs and horns, had ploughed ledges and cleaved fissures and fastened crags and pinnacles upon it, so that, while its surface was full of man-traps and blind ways, the human spider might still find some hold for his claws."[471]

In his article "Family of fallen climber follows trail of memories," *Denver Post* columnist Rich Tosches describes Blanca Peak: "Majestic and rugged… Blanca is as beautiful and forbidding a place as any in Colorado, a 14,345-foot tower of trees and rock and ice and snow and howling wind. It is a place where eagles fly."[472] And where some fall and die. In August 1995, Jim Mills found out just how imposing and dangerous Blanca Peak can be. While downclimbing a narrow, exposed route from the summit, he slipped.

Just like that, 43-year-old Mills (a family man with four kids) was gone. Tosches says search-and-rescue teams on foot and in helicopters searched for five days, but couldn't find a trace of the man.[473] "It is so rugged and jagged in that area that you could fall in a crevice and someone could walk 5 feet from you and just not see you," said Huerfano County Sheriff Bruce Newman, who worked in the department at the time of Mill's disappearance.[474]

---

470 John Fielder and Mark Pearson. *Colorado's Wilderness Areas*. Englewood, CO: Westcliffe Publishers, 1994, p. 238.

471 Lee Stetson (ed.). *The Wild Muir*. Yosemite National Park, California: 1994, p. 157.

472 Rich Tosches. "Family of fallen climber follows trail of memories." *The Denver Post*: 10/2/05, p. 1C.

473 Rich Tosches. "Family of fallen climber follows trail of memories." *The Denver Post*: 10/2/05, p. 1C.

474 Laurie Fox. "10-year mystery over disappearance put to rest." *The Dallas Morning News*: 10/16/05.

Tosches explains that Jim and his younger brother David started the climb together that morning. David turned back before noon, when they encountered snow and ice on the route, but Jim went on. David said, "When we hit the snow I had a gut feeling that it wasn't right, that it was the wrong day. Jim, though, he was always going to do what he set out to do."[475] "He was pretty much invincible in his mind," Mill's wife said. The two men were climbing Blanca from the more dangerous backside of the mountain.[476]

Bill Middlebrook, who runs the website 14ers.com, says there are at least two ways up Blanca from that direction. Both are dangerous. "Those are routes that I wouldn't do without a lot of time, a partner and rope," Middlebrook said. Jim Mills had neither in the end.[477] It seems to me many of these 14er fatalities boil down to the same critical, elemental fact: someone died because they didn't use good judgment.

David had enough sense to turn around on this difficult route when spotty conditions were encountered. Jim did not. Ten years later, on September 11, 2005, two climbers descending from the summit of Blanca Peak found Jim's body, finally bringing closure to Mill's family and friends.[478] A tragedy for sure, but a valuable lesson for the rest of us: skill, experience, and determination are useless without commonsense.

Despite the mountain's life-threatening temperament, Gerry Roach has good things to say about Blanca Peak in *Colorado's Fourteeners*: "Mighty Blanca carries many accolades. Blanca is Colorado's fourth highest peak, the highest peak in Colorado outside the Sawatch Range, the highest peak in the Sangre de Cristo Range and the highest peak in Alamosa, Costilla and Huerfano Counties... If you traveled south from Blanca, you would have to go all the way to the high volcanoes in central Mexico to find a higher peak... Soaring 7,000 feet above its valleys, Blanca carries its honors well. Climb it."[479]

I did, slowly and carefully, and on the summit was alone until the two climbers from Ellingwood Point arrived: an attractive and apparently cold young woman and her boyfriend. Yes, there was a nip or two in the air. We exchanged pleasantries and chitchatted a bit before descending from the summit together. The young woman and I talked most of the way down, to the chagrin of her boyfriend, and then said our good-byes as they proceeded to their campsite in the high basin above timberline and me to mine on the switchback down the mountain.

---

475 Rich Tosches. "Family of fallen climber follows trail of memories." *The Denver Post*: 10/2/05, p. 5C.
476 Laurie Fox. "10-year mystery over disappearance put to rest." *The Dallas Morning News*: 10/16/05.
477 www.14ers.com.
478 Laurie Fox. "10-year mystery over disappearance put to rest." *The Dallas Morning News*: 10/16/05.
479 Gerry Roach. *Colorado's Fourteeners*. Golden, CO: Fulcrum Publishing, 1999, p. 171.

July

Thirteen more 14er summits would feel the weight of my well-worn hiking boots during 2000, making it my most prolific fourteener season. Next up, over the long Fourth of July weekend, were some Mosquito Range peaks: Mount Shavano, Mount Princeton, and Mount Democrat, which fell on Saturday, Sunday, and Monday respectively. Mount Shavano (14,229 ft.) was named after a powerful Tabeguache war chief and medicine man, but his medicine couldn't stop bullets. He was shot by the father of a boy who died after one of his cures failed.[480]

Summitpost.org says the best way to climb Mount Shavano (14,229 ft.) is on the "Angel of Shavano" route. The Angel of Shavano is a distinctive snow formation appearing on the eastern flank of Mount Shavano each spring. One legend claims the snow signifies an Indian princess who, during a severe drought, knelt at the foot of the mountain and prayed for rain. Her tears, in the form of melting snow, provided valuable moisture to the Arkansas River Valley below. That's the only Shavano legend I'm privy to, so on to the summit.

From the Blank Gulch trailhead (adds Summitpost.org), the "Angel" climb is a 7.9 mile round-trip Class 2 route through 4,430 feet of elevation gain. At about 11,000 feet, before a series of switchbacks, leave the trail and bushwhack west, climbing up a drainage to the base of the "Angel with outstretched arms." An ice axe is recommended (even during the summer) on this snow route. At the top of the "body," continue climbing up an "arm." The northern arm is the preferred route as it leads directly toward the top. Once out of the arm, it's a talus scramble to the summit.[481]

I opted to bypass the Angel route, tackling Shavano entirely on the easier and faster Blank Gulch Trail instead, and after completing the climb, headed for Mount Princeton (14,197 ft.), the southernmost of the Collegiate Peaks and Colorado's 20th highest. This striking mountain stands by itself in the Sawatch Range between two U-shaped glacial valleys and dominates the Arkansas River Valley skyline west of Buena Vista.[482]

One legend claims the cliffs of Mount Princeton are the hiding place of a treasure stashed there centuries ago by Spanish raiders from New Mexico. The southern slopes of Princeton, where the treasure is allegedly hidden, are known as the Chalk Cliffs.[483] I skipped the cliffs and treasure hunting on this trip, and after car-camping on a switchback near tree line and climbing Princeton on July 2–and taking in some magnificent Arkansas Valley views from her summit–I

---

480 http://www.fourteenerworld.com/.
481 http://www.summitpost.org/show/mountain_link.pl/mountain_id/347.
482 http://www.summitpost.org/show/mountain_link.pl/mountain_id/205.
483 http://www.fourteenerworld.com/.

made a beeline for the Mosquito Range and Mount Democrat.

Park County is home to the locally popular and well-known Mosquito Range fourteeners, Mounts Democrat, Lincoln, and Bross. "Unlike other Colorado regions, Park County has no destination resorts," says Gary Nichols, the county's director of tourism and community development. "But dispersed recreation, like climbing the fourteeners, is a primary driver in our... economy; it represents an estimated 80 percent of seasonal income from tourist-dependent businesses."[484]

According to an August 2007 economic report issued by Colorado State University, the average climber spends $168 a day when tackling Mosquito Range fourteeners. These economic benefits are second only to those generated by such premium outdoor activities as river rafting.[485] Although I'm on the low end of fourteener climbing spending statistics, I was happy to contribute to Park County's coffers by climbing Mount Democrat on July 3. Democrat (14,148 ft.) is a Class 2 climb when tackled on the standard route starting at Kite Lake, and is Colorado's 29th highest peak.

The Kite Lake Trail starts at 12,000 feet on a gradual slope that steadily steepens as it climbs the saddle between Mount Democrat and Mount Cameron, where it starts to switchback. From there, head southwest up the east ridge of Mount Democrat to a false summit. The real one is a couple hundred yards farther and a hundred or so feet higher. After summiting Democrat, most climbers do an about-face and follow the same route down, unless they're climbing Mount Lincoln. If that's your plan, descend to the Democrat-Cameron saddle, then hike eastward and upward past Mount Cameron to the summit of Lincoln. From Mount Lincoln, nearby Mount Bross (14,172 ft.) is within easy (weather permitting) striking distance.

While climbing Democrat, a few other peak seekers and I were caught in a not-so-rare summer snow squall that reduced visibility to mere feet and turned us all around just short of the summit. We scrambled carefully back down the snow-covered mountain all the way to the Kite Lake trailhead, and within an hour, the storm blew over and there wasn't a cloud in the cobalt blue sky. What to do? Head for home or back up the mountain? I did another about-face and hiked-climbed Democrat again, all the way to the top this time, and upon summiting, was treated to spectacular views of snow-speckled peaks for as far as the eye can see.

I'd envisioned climbing Mount Democrat—my fourteenth fourteener—on the 4th of July, but settled for July 3 instead and returned home to spend America's birthday relaxing, for a change, in Colorado Springs. It was good to be home, especially

---

484 Trust for Public Land (TPL). "Taking the Initiative in the Mosquito Range." *Land & People*: Spring/Summer 2008, p. 40.
485 Trust for Public Land (TPL). "Taking the Initiative in the Mosquito Range." *Land & People*: Spring/Summer 2008, p. 40.

after a somewhat harrowing winter stormlike experience on Mount Democrat. The next weekend, I returned to Kite Lake and climbed Lincoln and Bross.

Mount Lincoln (14,286 ft.) is the 8th tallest mountain in Colorado and the highest point in Park County. It's usually climbed after Mount Democrat and before Mount Bross, as many climbers bag all three peaks together. Individually, each of them is a relatively easy hike-climb, but doing the triumvirate does require some effort. From the Democrat-Cameron saddle, it's a Class 2 stroll past Cameron and up Lincoln's west ridge to the summit. Mount Lincoln was named after President Abraham Lincoln (a Republican) to get back at the Democrats for naming its sister peak Mount Democrat.[486]

Believed to be the highest peak in Colorado at the time, Mount Lincoln was so christened in 1861. When President Lincoln heard the news, what do you suppose he did? He sent his friend Speaker of the House Schuyler Colfax west on a mission to thank the people of Colorado.[487] On a related note, one of Colorado's less well-known fourteeners is named after the man who served as Lincoln's secretary of war: Mount Cameron (14,238 ft.) was so named for Simon Cameron (1799-1889). Mr. Cameron was Abraham Lincoln's first secretary of war, a rival he hired and then fired for corruption after less than 10 months in office.

Because Mount Cameron lacks saddle prominence from Mount Lincoln, it's usually omitted from the official list of Colorado's fourteeners. Regardless, in 2006, the Trust for Public Land acquired 130 acres of mining claims in the Mosquito Range, including the summit of Mount Cameron.[488] From Cameron and Lincoln, it's a hop, skip, and a stumble (if not careful) over to Mount Bross (14,172 ft.). The mountain was named after the (then) lieutenant governor of Illinois, William Bross, who owned mining property in nearby Alma, Colorado.[489]

Mount Bross is the 22nd highest mountain in Colorado and an easy Class 1 hike from Mount Lincoln. The most popular trailhead for both peaks is Kite Lake. If you climb Bross from the Kite Lake trailhead, skipping Democrat and Lincoln, it's a 2.8-mile jaunt through 2,170 feet of elevation gain. The summit is somewhat nondescript (just an oversized, gently rolling mountaintop), but they don't all have to be lung-busting hard and drop-dead beautiful, even though that's the norm (beautiful), if you ask me.

Rock "bunkers" built to protect resting climbers from the wind are about the only indication that you're actually on the summit of Mount Bross proper, but

486 http://www.summitpost.org/show/mountain_link.pl/mountain_id/20.
487 http://www.fourteenerworld.com/Trivia/TOD20060928.htm.
488 Trust for Public Land (TPL). "Taking the Initiative in the Mosquito Range." *Land & People*: Spring/Summer 2008, p. 40.
489 http://www.fourteenerworld.com/.

knocking off two or more fourteeners in a day (no matter how supposedly easy) is always a rewarding peak bagging bonus. The next weekend I was back in the Sawatch Range, west of Buena Vista, to climb Mount Yale. Yale (14,202 ft.) is the 8th tallest peak in the Sawatch Range and 21st highest in Colorado.

Summitpost.org notes that Mount Yale received its name from J. D. Whitney (after his alma mater) in 1869. Whitney (head of the Harvard School of Mining), along with W. H. Brewer, C. F. Hoffman, and students from the school's first graduating class came to Colorado to survey peaks for two reasons: to give the students some practical field experience and to investigate rumors that Mounts Harvard, Yale, Princeton, and Columbia were higher than the highest peaks in California.[490] Another morsel of Mount Yale history occurred in 1956.

That winter, a cantankerous runaway horse named "Bugs" got stranded on the Continental Divide next to Yale and was kept alive until spring with airdrops of hay and oats. The news media dubbed the horse "Elijah," after the biblical prophet who was kept alive by ravens in the wilderness.[491] Fourteenerworld. com says, "In the winter of 1956-57, the eyes of the animal loving world were focused on a 12,500-foot saddle between Harvard and Yale where an old bay horse was spending the winter. People around the world sent money to help finance Operation Haylift, through which planes dropped tons and tons of hay to the horse."[492]

At least Elijah had a nice view while waiting for spring to arrive, because according to the Colorado Fourteeners Initiative, "Mount Yale rises dramatically from the surrounding valleys and offers a stunning view of 30 fourteeners from its summit."[493] A somewhat similar, but more recent incident (with a llama) occurred on Pikes Peak, as detailed in this *Associated Press* story.

"A lone llama wandering near the summit of Pikes Peak for a month has been captured and is heading to a new home. Tracy Ducharme and Mike Shealy, both of Black Forest, Colo., trekked up the 14,110-foot mountain Friday to find the little white beast of burden. They took two llamas with them, hoping Homer's herd instincts would lure him to them.[494]

"The two split up and Ducharme spotted the llama, which bounded after her llama, Dancer. She then slipped a rope around his neck. 'I

490 http://www.summitpost.org/show/mountain_link.pl/mountain_id/311.

491 Walter R. Borneman, Lyndon J. Lampert. *A Climbing Guide to Colorado's Fourteeners* (updated 2nd edition). Boulder, Colorado: Pruett Publishing, p. 104-105.

492 http://www.fourteenerworld.com/.

493 Colorado Fourteeners Initiative (CFI). "2008 Peak Projects." *Fourteener:* Spring 2008, p. 5.

494 Associated Press (AP). "Lone llama rescued after month on Pikes Peak." *AP:* 10/2/09.

2000: PEAK YEAR ❧

dubbed him Homer because of his little odyssey,' she said, referring to the classic Greek novel *Odyssey* by the poet Homer. llamas are domesticated animals and don't have the instincts to survive in the wild.[495]

"Rescuers speculate that the llama might have escaped from a stock trailer on the plains below or from hikers who were using it as a pack animal. But no one has reported a missing llama. Riders on a cog railway on the mountain first reported seeing the llama about a month ago. It roamed the mountain's south slope, living off alpine vegetation while trying unsuccessfully to make friends with a herd of bighorn sheep."[496]

After climbing Mount Yale on July 15 and spotting no wayward horses or llamas along the way, I took a break from 14ers to visit the north rim of the Grand Canyon in Arizona and Cedar Breaks National Monument in Utah after attending a work conference in Las Vegas. Although Vegas is an interesting place to visit and a near-sensory overload for first-timers, it's not where one goes to "get away from it all." On the other hand, if you're stuck there, you might as well make the best of it.

Backpacker magazine has some Las Vegas hiking advice for wayward outdoorsmen and women: "Begin this tour after sunset—when the temperature shifts from broil to bake—and hike north to the Stratosphere, the tallest tower west of the Mississippi. Swing southwest to the Wynn Monument and its multitier waterfalls (fed by the Colorado River!). Continue past Venetian buttes, rocky skyscrapers, and ponds with flocks of flamingos—and don't miss the always faithful Bellagio geysers... No permits required, no overnight camping allowed."[497]

The state of Nevada, according to the National Parks Conservation Association (NPCA), is "often thought of as an endless desert, but this belies the fact that it's home to more mountain ranges than any other state." Great Basin National Park, for example, is a stellar sample of the state's rugged, mountainous terrain and natural beauty. Protected within the park's boundaries are towering peaks, glacier-studded valleys, fast-running creeks, sparkling alpine lakes, a variety of cave formations, expansive desert habitats, and ancient bristlecone pines (the world's oldest living trees).[498]

This remote desert park in east-central Nevada has (notes Reid Bramblett, in "America's Most Underrated National Parks") groves of gnarled bristlecone clocking in at more than 4,000 years old, as well as aspen, jackrabbits, and alpine wildflowers spread out over 77,000 acres that range from the basin floor

---

495 Associated Press (AP). "Lone llama rescued after month on Pikes Peak." *AP*: 10/2/09.
496 Associated Press (AP). "Lone llama rescued after month on Pikes Peak." *AP*: 10/2/09.
497 Backpacker. "The Vegas Strip, NV." *Backpacker*: 5/20/10, p. 49.
498 National Parks Conservation Association (NPCA). "Great Basin National Park." *Parks Lines*: 5/7/09.

at 5,000 feet to peaks topping 13,000 feet. The skies over isolated Great Basin rank among the darkest in the Lower 48 states, providing some of the best stargazing in the nation. In addition to mountains and stars, there are the glittering marble caverns of Lehman Caves and the 3.4-mile round-trip hike to Lexington Arch, a rare aboveground limestone arch that, at six stories high, is one of the largest in the United States.[499]

After exploring the wilds of Las Vegas, southern Utah, and northern Arizona, I returned to Colorado and during August, I completed the rest of the Collegiate Peaks. The Collegiates are all located in the Collegiate Peaks Wilderness, which protects into perpetuity eight summits exceeding 14,000 feet, including Colorado's 3rd and 5th highest, Mount Harvard and La Plata Peak, respectively. In addition, five peaks soar over 13,800 feet.[500]

Some of America's most picturesque rock, ice, and timberline wildlands are found in places like the high altitude Collegiate Peaks Wilderness, which straddles Colorado's Continental Divide in the Sawatch Range. This mountain range stretches nearly 100 miles, starting at Tennessee Pass near Vail and ending in the Marshall Pass area southwest of Salida. The Sawatch Range is home to 15 fourteeners and 14 centennial thirteeners, and no other Colorado mountain range contains as many peaks above 13,800 feet.[501]

According to the Colorado Mountain Club, if you want to show visitors what Colorado mountains are all about, "Look no farther than the Collegiate Peaks. Looming above the Arkansas River valley like titans, mounts Harvard, Yale, Princeton and the rest form what's said to be the highest-average-elevation wilderness in the state. With eight fourteeners—including Colorado's third- and fifth-tallest peaks (Mount Harvard at 14,420 feet and La Plata Peak at 14,336 feet)—and five more summits that top 13,800 feet, the Collegiates are big-mountain scenery at its best."[502]

The Collegiate Peaks' 167,414 wilderness acres were set aside by Congress in 1980, and they're the fifth largest Colorado wilderness area behind the Weminuche (488,210 acres), Flat Tops (235,214 acres), Sangre de Cristo (186,368), and Maroon Bells (181,512), all of which I've been privileged to visit over the years. But due to the abundance of cherrystems, you're never more than five miles from a road in the Collegiates. Congress drew the area's odd boundaries to avoid blocks of private patented mining claims and historical mining activity.[503]

---

499 Reid Bramblett. "America's Most Underrated National Parks." *Yahoo! Travel*: 4/3/10.

500 John Fielder & Mark Pearson. *Colorado's Wilderness Areas*. Englewood, CO: Westcliffe Publishers, 1994, p. 162.

501 http://www.summitpost.org/show/mountain_link.pl/mountain_id/311.

502 Colorado Mountain Club (CMC). "Collegiate Peaks Wilderness." *Trail & Timberline*: Spring 2010, p. 25.

503 John Fielder and Mark Pearson. *Colorado's Wilderness Areas*. Englewood, CO: Westcliffe Publishers, 1994 p. 162.

Despite the cherrystems that make this wilderness far less "remote" than it should be, within the Collegiate Peaks Wilderness boundaries there are more fourteeners and high thirteeners than in any other wilderness area in the Lower 48 states. The wilderness runs along a 40-mile stretch of the Continental Divide and has the highest average elevation (as mentioned above) of any wilderness area in the contiguous United States. Its eight peaks over 14,000 feet include:[504]

- Mt. Harvard (14,420 ft.)
- La Plata Peak (14,340 ft.)
- Mt. Belford (14,204 ft.)
- Mt. Yale (14,202 ft.)
- Mt. Oxford (14,160 ft.)
- Mt. Columbia (14,073 ft.)
- Missouri Mountain (14,067 ft.)
- Huron Peak (14,012 ft.)

The wilderness area's five 13,800-foot-plus centennial thirteeners include:

- Grizzly Peak (13,995 ft.)
- Ice Mountain (13,951 ft.)
- Emerald Peak (13,904 ft.)
- North Apostle (13,860 ft.)
- Iowa Peak (13,831 ft.)

Like wilderness areas everywhere, the Collegiate Peaks is a place of spectacular beauty and would-be danger, and one where famed mountaineer Reinhold Messner would likely feel right at home. Although possibly less striking than his Dolomites and Alps, I'm certain Reinhold would find much to admire in the Collegiates, especially given his mountain-goat-like abilities and exploits. Among Messner's many mountaineering accomplishments, he literally reinvented climbing with his mid-seventies introduction of fast and light alpine-style techniques on high elevation routes.

Messner put these techniques to the ultimate test on Mount Everest. In *Climbing* magazine, Messner said, "What makes Everest so dangerous is not the steepness of its flanks, nor the sheer masses of rock and ice that can break off without warning; what is far worse is the reduced air pressure in its upper regions. This saps your judgment and strength, even your ability to feel anything

---

504 http://www.summitpost.org/show/mountain_link.pl/mountain_id/311.

at all. That's what makes you so vulnerable and afraid up there."[505]

In "Reinhold Don't Care What You Think," *Outside* magazine contributor Brad Wetzler explains that Messner used no oxygen, fixed ropes, established camps, or support teams—just a clean break from the siege tactics and mentality that had dominated the game until then.[506] At the top of Mount Everest, atmospheric pressure is 33 percent of that at sea level. Mountaineers who summit find 66 percent less oxygen available for breathing. That's not enough air to burn kerosene, not enough for a helicopter to take flight, and for some, it's not enough to live, but it was for Reinhold.[507]

> "Altitude is the great equalizer."
> —Author Unknown

Most climbers say Messner's most infamous feat was his solo ascent of Mount Everest, when he climbed the mountain's North Face carrying everything he needed on his back, returning to Base Camp in an astonishing four days. "It was like landing on the moon," says Conrad Anker, an Everest veteran from Bozeman, Montana. "After that, everything kind of pales in comparison." Tom Hornbein, the 70-plus-year-old U.S. climbing legend, adds, "Reinhold Messner really is a great tour de force. He's one of those people who shake paradigms."[508]

After a lifetime of paradigm-toppling climbing adventures, Messner has turned his attention to protecting and preserving the remaining wild regions of the world. "Deserts and mountains are a catalyst for our humanity," he says. "On them we can discover our human abilities and limitations. Nature, in the form of rugged scenery, is the best mirror of our souls."[509] Well said, Reinhold, for a European, but the modern-day idea of protecting an area in its primitive state as "wilderness"—wild land left as it is for its own sake—is an American construct.

During the mid-nineteenth century, a small but vocal group of explorers, trappers, hunters, anglers, artists, and writers led the charge for wilderness in the U.S., with Henry David Thoreau arguing from his pond-sized home in Concord, Massachusetts, that wilderness sanctuaries are a necessary complement to civilization. In "Wilderness: America's Lands Apart," *National Geographic* contributor John G. Mitchell explains that it was no easy task changing minds and

---

505 Robert Sullivan and Robert Andreas (eds.). *The Greatest Adventures of All Time*. Des Moines, IA: LIFE Books, 2000, p. 111.

506 Brad Wetzler. "Reinhold Don't Care What You Think." *Outside* online: 10/4/02, p. 3.

507 Robert Sullivan and Robert Andreas (eds.). *The Greatest Adventures of All Time*. Des Moines, IA: LIFE Books, 2000, p. 111.

508 Brad Wetzler. "Reinhold Don't Care What You Think." *Outside* online: 10/4/02, p. 4.

509 Reinhold Messner. *Free Spirit: A Climber's Life*. Seattle, WA: The Mountaineers, 1998, p. 261.

convincing the masses to support wilderness protection in those days. It held little allure for early settlers; deep and dark, full of beasts and demons; for most, it was good for only one thing: taming.[510]

But very slowly, as America's vast blanket of wildlands was reduced to ever-smaller island remnants, this frontier mentality gradually transformed and wilderness became a refuge—for both wildlife and people. As a nation, we began to realize that, like Aldo Leopold said, "Wilderness is a resource which can shrink but cannot grow," and (I would add), we are still a great nation mostly because of the wild places and wide-open spaces that still exist to collectively morph and mold our shared national character.[511]

According to David Jenkins, REP's government affairs director, "In wilderness liberty is found in its most fundamental form. When people lose touch with the self-reliance, freedom and grandeur embodied in wilderness, the most basic foundations of democracy are undermined." American democracy, David adds, owes its existence and longevity as much to the vast wilderness landscapes that greeted our forefathers as it does to the Declaration of Independence and Constitution.[512]

A century after signing the Declaration of Independence (and 90 years before the Wilderness Act was passed), Americans launched a second revolution that changed the course of human and wildlife history: the establishment of Yellowstone National Park as the world's first national park. Today it's easy to take national parks for granted because they're popular and widespread, but in 1872, the "national park idea" was a radical one. It challenged the then generally accepted Euro-American notion that nature was foreboding, dangerous, and ungodly.

For a historical perspective, consider that General George Armstrong Custer was soundly defeated by free-roaming Native Americans at the Little Bighorn (Montana) in mid-1876. At that time, visiting our one and only national park would certainly have been a grand adventure. And while the National Forest System was created in 1897, the National Park Service wasn't created until 1916. The army managed Yellowstone until then.[513]

According to wilderness historian Roderick Nash, today's appreciation of wilderness represents one of the most remarkable intellectual revolutions in the history of human thought about the land. Wilderness has evolved from "an earthly hell to a peaceful sanctuary where happy visitors can join the likes of John Muir and John Denver in drawing nearer to divinity." Such a perspective would

510 John G. Mitchell. "Wilderness: America's Lands Apart." *National Geographic*: November, 1998, p. 3.
511 David A. Lien. "Silent Sentinels." *Association of Air Force Missileers (AAFM) Newsletter*: March 2009, p. 1.
512 David Jenkins. "The Wild Side of Conservation." *C.E.P. Quarterly*: Summer 2005, p. 8.
513 Rick Hartman. "#22 The Wilderness Act." *Apex to Zenith*: 2nd Quarter 2008, p. 24.

have been absolutely incomprehensible to, say, a Puritan in New England in the 1650s.[514]

In protecting Yellowstone (1872), then Yosemite (1890) and General Grant (now part of Kings Canyon) and Sequoia National Parks (also in 1890), Americans started turning their backs on the wayward belief that wilderness is something to be feared, tamed, and exploited. Instead, they decided the natural world and wilderness are worth preserving in their own right. Much has been accomplished since 1872—more national parks have been set aside and wilderness areas protected—but since then, even more wildlands have been lost, and today there are many, countless more threats to our public lands. Take roads for instance.

The Thorofare Ranger Station in the southeast corner of Yellowstone National Park is right in the heart of what mapping experts say is the most remote place in the contiguous United States. *Denver Post* contributor Robert W. Black explains (in "Wyo. Ranger station is far out—way out") that it was pegged as the most remote place by Susan Boswell, president of Cartographic Technologies. Her company was asked by an automaker to find the place most distant from a publicly maintained road. Boswell found that roads are "uncomfortably close" nearly everywhere, and the nearest road to the Thorofare Ranger Station is only 20 miles away.[515]

"There's nothing around, but in terms of actual distance as the crow flies, you're not too far from a house or from people or from roads," Boswell said. The second-most remote spot is in the Bob Marshall Wilderness Area in Montana, 18 miles from a road. The third is in Idaho's Frank Church River of No Return Wilderness, 16 miles from a road.[516] Considering that I've personally hiked over 28 miles through 5,350 feet of elevation gain in a single day (to Kings Peak, and back, in Utah's 456,705-acre High Uintas Wilderness), I can tell you from personal experience that being 20 miles from a road isn't very far.[517]

Despite its "Nearby Faraway" (the title of a book by David Petersen) proximity to roads, the Greater Yellowstone region still supports one of the world's few remaining fully functioning ecosystems. As told by *Backpacker* magazine contributor Jim Gorman, in "Where Rangers Hike," Orville Bach has been a seasonal ranger in Yellowstone for 32 years and has some incredible stories about his treks in the park. Like the time he was canoeing on Yellowstone Lake <u>and came within</u> feet of a cougar swimming between Breeze and Pumice points,

514 John G. Mitchell. "Wilderness: America's Lands Apart." *National Geographic*: November, 1998, p. 32.

515 Robert W. Black. "Wyo. Ranger station is far out—way out." *The Denver Post*: 7/14/00.

516 Robert W. Black. "Wyo. Ranger station is far out—way out." *The Denver Post*: 7/14/00.

517 Kings Peak, the highest point in Utah (13,528 ft.), was my 33rd state highpoint. I hiked-climbed it on the Henrys Fork Trail starting at 2:00 a.m., summited at 9:50 a.m., and returned to the trailhead at 4:30 p.m., for a 14½-hour round-trip covering 28½ miles through 5,350 ft. of elevation gain. My previous one-day hiking record was 26 miles through 7,510 feet of elevation gain (on the Barr Trail) on Pikes Peak in Colorado.

or the cave in Hayden Valley that held the skeletons of seven buffalo, dragged there by a dining grizzly.[518]

One region Ranger Bach returns to time and again is the Thorofare District below Yellowstone Lake. During 2004, only 315 people journeyed into the Thorofare. "That's the wildest," he says. "It's backed by the big Absaroka Range peaks, like Table and Trident. It's teeming with elk, moose, and lots of grizzlies and wolves."[519] It's a land of towering waterfalls, dark-timbered forests and snow-capped mountains, and it is some of the wildest country left in the contiguous United States and the world. During September 2008, I backpacked 65 miles over 6 days into the Thorofare Valley, encountering both a wolf and grizz (at close range!) along the way, but that's a story for another time.[520]

One evening while camped there (explains Gorman), Ranger Bach heard an entirely new sound. "We had cliffs to either side of us, and up starts this chorus of wolf howls, reverberating back and forth in the valley," he recalls. Wolves were reintroduced to the park in 1995, but this was the first time Bach had made their acquaintance. "That was when I realized Yellowstone was whole again," says Bach.[521] Today, the Greater Yellowstone region, with Yellowstone National Park at its wild core, is the largest intact ecosystem in the Lower 48, and one of only a handful in the country and world that hasn't been decimated by the hands of man.

Celebrated documentarian Ken Burns knows better than most the meaning and importance of national parks. "They are America's best idea," he says. "First, you write a Declaration of Independence. Then you have Thoreau and Emerson writing about nature. They're essentially emulating Jefferson, but taking him a step further by saying, 'To be free, I also need to be in nature.' What's the next impulse? To save it. That creates conflict, but it's a magnificent tension... In the face of these magnificent landscapes, which are so different from what they know, where they're reminded of God's handiwork, they find themselves."[522]

"Our European ancestors essentially lived a geographically proscribed life, rarely venturing beyond where they were, and all of a sudden, the combination of land and democracy set in motion one hell of a great story," Burns adds. "We think the best one of all is the story of how a fledgling democracy suddenly decided you could set aside large tracts of natural land, not for the kings and royalty and the very rich who had normally cornered the market on beautiful

---

518 Jim Gorman. "Where Rangers Hike." *Backpacker:* December 2005, p. 88.

519 Jim Gorman. "Where Rangers Hike." *Backpacker:* December 2005, p. 88.

520 See article: "Lien treks in Yellowstone." Grand Rapids *Herald-Review:* 10/1/08, p. 3c. Also see: "Grand Rapids adventurer will share stories at Grand Rapids Area Library." Grand Rapids *Herald-Review:* 6/28/09, p. 10.

521 Jim Gorman. "Where Rangers Hike." *Backpacker:* December 2005, p. 88.

522 Tracy Ross. "Ken Hits The Trail." *Backpacker:* December 2005, p. 20.

places, but for everybody for all time."[523]

In Yellowstone's Lamar Valley (notes *Associated Press* contributor Matthew Brown), Burns watched wolves, a pack of a dozen pups and adults, circling the edge of a bison herd on the lookout for winter-weakened prey. Using a powerful spotting scope, Burns saw the animals crane their necks skyward seconds before their howls spilled across the valley floor. "Here it comes," Burns said, raising a hand for quiet just before the sound reached their ears. The filmmaker let out his own howl in response, then offered a quote from one of the characters in his series: "Now let me die, because I am happy."[524]

Famed writer Wallace Stegner said, "National parks are the best idea we ever had. Absolutely American, absolutely democratic, they reflect us at our best rather than our worst." This quintessential American idea ignited a worldwide national parks movement, and today, we can visit over 1,200 national parks in 100 countries across the globe, including 58 in the United States (as of 2003) covering almost 52 million acres.

When the first World Parks Congress convened in 1962, there were not quite a million square miles of protected lands around the globe. During its 2003 meeting in South Africa, the fifth World Parks Congress counted some seven million square miles of protected lands and waters worldwide, a quarter of it within national parks and designated natural areas. Today, almost half of Greenland is protected, a third of New Zealand, and a tenth of Italy. Our greatest idea is catching on, fast.[525]

Although no national parks owe their preservation to George W. Bush, many of today's wild places owe theirs to Teddy Roosevelt. On a cool, cloudy day in September 1901, Vice President Theodore Roosevelt was hiking in the Adirondacks of New York. As Edmund Morris describes in his TR biography, *Theodore Rex*, "The clouds unexpectedly parted, sunshine poured down on his head, and for a few minutes a world of trees and mountains and sparkling water lay all around stretching to infinity."[526]

There could not have been more auspicious surroundings for the great hunter-conservationist, who hours later learned he had become (at only 42-years-old) America's 26th (and youngest) president. Theodore Roosevelt walked off the mountain and over the next seven years and sixty-nine days his decisive actions in defense of wildlands and wildlife bequeathed to our nation and world an unmatched conservation legacy that has delivered incalculable benefits to all. On that fateful day

---

523 Scott Kirkwood. "National Parks: The Film." *National Parks*: Spring 2009.

524 Matthew Brown. "Famed documentarian tells story of Yellowstone, other parks." *Associated Press*: 2/2/09.

525 John G. Mitchell. "Our Great Estate." *Sierra*: March/April 2004, p. 30.

526 REP America. *Conservation is Conservation*. Albuquerque, NM: REP Environmental Educational Foundation, 2001, p. 12.

in 1901, a cultural revolution dawned on the United States of America.[527]

As told by *Wild Earth* contributor Curt Meine, in "Conservation and the Progressive Movement," TR's revolution challenged the assumption that had dominated national development for generations: that the earth was merely a storehouse of inexhaustible resources made solely for the indulgence of the present generation's most privileged species.[528] President Roosevelt started off by protecting a small piece of land in Florida. In setting aside the first national wildlife refuge (Pelican Island) in 1903, TR preserved a patch of America that is now the smallest of this nation's formally protected lands—a mere five acres.

Roosevelt's life on the storm-tossed plains of the Dakota Territory (1883-1887) running cattle and hunting big game taught him the value of conservation, notes REP policy director Jim DiPeso. That formative experience led to his insight that protecting lands for the benefit of all, present and future, was an exercise in democracy and essential to keeping America strong. TR spent the spring of 1903 traveling the West and pounding that lesson home.[529]

He called the preservation of Yellowstone "essential democracy." On California's central coast, he scolded townspeople for vandalizing a stately redwood with trifling advertising posters. At the rim of the Grand Canyon, he implored his countrymen to "leave it as it is. You cannot improve upon it. The ages have been at work on it and man can only mar it." And in Yosemite National Park, the crown jewel of the Sierra, DiPeso adds, TR tramped through the granite wilderness and sequoia groves with John Muir like a schoolboy on summer break, yelling "Bully!" upon waking up at camp on Glacier Point to find it covered with a late-season snow.[530]

National Parks contributor Seth Shteir explains (in "To Dare Mighty Things") that Roosevelt's administration doubled the number of national parks from five to 10, protecting the majestic blue waters of Crater Lake National Park, the rich archaeological resources of Mesa Verde, and the caverns of Wind Cave National Park. After a camping trip with the sure-footed John Muir in Yosemite National Park, Roosevelt added the stunning Yosemite Valley and Mariposa Grove to the park. In all, he protected 243 million acres of land, including 150 national forests, 51 bird sanctuaries, and 18 national monuments.[531]

One of the greatest gifts to future generations that came out of TR's far-sighted conservation efforts and the environmental movement of the 1950s and

---

527 REP America. *Conservation is Conservation.* Albuquerque, NM: REP Environmental Educational Foundation, 2001, p. 12.

528 Curt Meine. "Conservation and the Progressive Movement." *Wild Earth*: Summer/Fall 2003, p. 59.

529 Jim DiPeso. "Teddy Roosevelt was a Backpacking Elitist Too." *FrumForum*: 10/28/10.

530 Jim DiPeso. "Teddy Roosevelt was a Backpacking Elitist Too." *FrumForum*: 10/28/10.

531 Seth Shteir. "To Dare Mighty Things." *National Parks*: Fall 2008.

1960s that followed was the Wilderness Act of 1964. This act recognized the value to the nation and humanity of leaving some lands untouched and primeval. It was a sea-change, a paradigm shift in human thought about our relationship with the land.

On August 20, 1964, Senator Clinton P. Anderson (D-NM) said of the proposed Wilderness Act: "In the age of automation, mechanization, and exploitation of our vast natural resources, the amount of public lands shielded from the onslaught of man's ambition and genius becomes [ever] smaller. Our task in this age has been to stand off and ponder the consequences of that onslaught. I believe that this bill contains our verdict, and I believe that we can all be grateful that the verdict came while we still had wilderness to preserve."[532]

Senator Frank Church (D-ID) adds, "We westerners have known the wilds during our lifetimes, and we must see to it that our grandchildren are not denied the same rich experience during theirs."[533] Senator Hubert Humphrey (D-MN) formally proposed the Wilderness Act on the Senate Floor in 1956. Almost a decade later, on September 3, 1964, after eight years of deliberations, 18 congressional hearings, and 66 drafts, an act creating the National Wilderness Preservation System passed under the pen of President Lyndon B. Johnson, protecting into perpetuity 9.1 million acres of public land in 13 states.[534]

Wilderness Act architect Howard Zahniser said, "Out of the wilderness, has come the substance of our culture, and with a living wilderness... we shall have also a vibrant, vital culture, an enduring civilization of healthful, happy people who... perpetually renew themselves in contact with the earth. We are not fighting progress, we are making it." Zahniser wrote the first draft of the Wilderness Act in 1956 and steered it through 65 rewrites and 18 public hearings, but he died during May 1964, just four months before it was signed into law.[535]

In a fitting tribute ("America has lobbyist to thank for wilderness") to Zahniser, Doug Scott (author of *The Enduring Wilderness*) explains that from the wild blessings of 3,700 acres of quiet swamp in suburban New Jersey, just 26 miles from Times Square, to the stunning San Gabriel Wilderness on the rim of Los Angeles, to a single road-free sweep of Alaska's Brooks Range the size of more than six Yellowstones, Congress has used the Wilderness Act to single out diverse, wild places for the strongest form of land protection available.[536]

Colorado Backcountry Hunter and Angler (BHA) Bill Sustrich hit the nail on the head when he said, "From my own observations, I have seen nothing yet

532 Doug Scott. *The Enduring Wilderness*. Golden, Colorado: Fulcrum Publishing, 2004, p. 4.
533 Doug Scott. *The Enduring Wilderness*. Golden, Colorado: Fulcrum Publishing, 2004, p. 51.
534 John G. Mitchell. "Wilderness: America's Lands Apart." *National Geographic*: November, 1998, p. 12.
535 http://www.wilderness.net/index.cfm?fuse=feature0504.
536 Doug Scott. "America has lobbyist to thank for wilderness." *Desert Morning News*: 2/26/06.

created by mankind that offers the degree of habitat protection that is achieved through wilderness designation."[537] Wilderness designation is the Holy Grail for hunters, anglers, hikers, backpackers, climbers, equestrians, outfitters, and outdoor enthusiasts everywhere, the strongest possible protection public land and wildlife habitat can have bestowed upon it. As Bill's hunting brethren Theodore Roosevelt said, "The wildlife and its habitat cannot speak. So we must and we will."[538]

And fighting for wilderness is what TR would still be doing today if he was around, and he'd be inspired by the fact that the Wilderness Act now protects some of the wildest parts of our federal lands in 44 states, a sort of Smithsonian Institution of the original American landscape (says Doug Scott) and the gold standard for wildlife habitat and hunting grounds everywhere (I say), but this wild Smithsonian not only preserves our past, it shines a ray of hope toward the future.[539] It says to those generations yet to come that we are not just a short-sighted, obese, overpopulated society of increasingly greedy consumers, flabby motorized wreckreationists, and brain-dead television watchers.

In some of the most poetic language ever penned by U.S. legislators, Congress defined wilderness as "an area where the earth and its community of life are untrammeled by man, where man himself is a visitor who does not remain. In order to assure that an increasing population, accompanied by expanding settlement and growing mechanization, does not occupy and modify all areas within the United States... it is hereby declared to be the policy of the Congress to secure for the American people of present and future generations the benefits of an enduring resource of wilderness."

In "Forty Years of Wilderness Progress," Utah conservationist Gale Dick explains that a congressionally designated wilderness is to have no permanent development, human habitation, commercial enterprise, permanent or temporary roads, motor vehicles or equipment, nor any motorboats or landing of aircraft, except where such uses have already been established.[540]

Colorado outfitter Gavin Selway (in "Wilderness is good for business") adds: "Rescue in wilderness is not a problem. I have a cell and a satellite phone and can call for help if I need to. Emergency access is allowed in wilderness, by foot, by vehicle and by aircraft as necessary. Our clients pay for a backcountry experience on horseback. They don't want to see ATVs and bicycles. The horses are safer without those distractions as well, although as a responsible outfitter, I

---

537 Bill Sustrich. "Browns Canyon." *Landscapes:* December 2009, p. 5.
538 Rick Hartman. "#22 The Wilderness Act." *Apex to Zenith:* 2nd Quarter 2008, p. 25.
539 Doug Scott. "America has lobbyist to thank for wilderness." *Desert Morning News:* 2/26/06.
540 Gale Dick. "Forty Years of Wilderness Progress." *Save Our Canyons:* Spring 2004, p. 2.

make sure my horses are as safe as possible in every condition."[541]

Despite what some people say, wilderness does not "lock up" public lands; it "frees" them from the incessant pressures of commercial development and preserves irreplaceable wildlife habitat and hunting grounds.[542] For example, the following activities are allowed in wilderness areas:

- Hunting (except in national park wilderness) and fishing
- Hiking, backpacking, and camping
- Float trips, canoeing, kayaking
- Horseback riding and pack trips
- Outfitting and guiding
- Wheelchairs (including certain motorized wheelchairs)
- Scientific research and nature study
- Control of fire, and insect, and disease outbreaks
- Livestock grazing and related facilities, where previously established
- Mining on preexisting mining claims
- Continued use of tracts of private or state land within the boundaries of some wilderness areas, with reasonable access.[543]

These activities are prohibited in wilderness areas:

- Road building
- Oil and gas drilling
- Logging
- Mechanical vehicles such as dirt bikes and off-road vehicles (certain motorized wheelchairs are allowed)
- New mining claims
- New reservoirs, power lines, and pipelines.[544]

Wildernesses areas are established by Congress on any public land administered by a federal agency, and over 40 percent of our nation's wilderness acres are found in national parks. Within the National Park System, there are 45 park units that contain designated wildernesses, comprising 43 million acres of our public lands. There are also nearly two-dozen parks with lands recommended for wilderness status that are awaiting congressional approval.

As of 2005, congressionally designated wildernesses accounted for just 4.7

---

541 Gavin Selway, Bearcat Stables. "Wilderness is good for business." *Summit Daily News:* 8/25/10.

542 Gale Dick. "Forty Years of Wilderness Progress." *Save Our Canyons:* Spring 2004, p. 2.

543 Save Our Canyons (SOC). "Introduction to Wilderness." *SOC:* Autumn 2008, p. 6.

544 Save Our Canyons (SOC). "Introduction to Wilderness." *SOC:* Autumn 2008, p. 6.

percent of all land in the United States. Outside of Alaska, only 2.5 percent is protected, an area about the size of South Dakota. In my boyhood home state of Minnesota, there is (out of a total land base of approximately 50.9 million acres) 816,268 acres of wilderness in three wilderness areas: the Boundary Waters Canoe Area Wilderness (810,088 acres), the Agassiz Wilderness (4,000 acres), and the Tamarac Wilderness (2,180 acres).

Is 1.6 percent of Minnesota's landmass in three wilderness areas adequate for all the other species that depend on wild habitat? What about 2.5 percent in the Lower 48? In 1964, the Wilderness Act established 54 wildernesses in national forests across 13 states and decreed that 9.1 million acres were to be forever protected in their natural state. Since then, approximately 97 million additional acres have been added, with more than half of that total (57.5 million acres) in Alaska.

Only Connecticut, Delaware, Iowa, Kansas, Maryland, and Rhode Island have no protected wilderness. The largest wilderness area is the Wrangell-Saint Elias in Alaska (9 million acres), and the smallest, Pelican Island, Florida (5 acres). The largest combined wilderness in the Lower 48 is the neighboring Frank Church-River of No Return and Gospel-Hump Wildernesses, totaling 2.6 million acres.[545]

Ronald Reagan signed bills creating 44 different wilderness areas (covering 10.6 million acres in 31 states), the most of any president. Jimmy Carter created the most wilderness acres, 66.3 million, most of it in Alaska. As of early 2004, President Bush had signed legislation creating only four new wildernesses.[546] Indeed, as Jim DiPeso (policy director for REP America) said, "The [Bush] Interior Department is in the process of expunging the word 'wilderness' from the dictionary."[547]

Whether we drink in luminous wilderness scenery during our summer vacations or live within hiking or weekend driving distance, these unique backcountry havens provide a wild and necessary backdrop to our civilization-stressed and -strained lives. Doug Scott (the Campaign for America's Wilderness policy director) says they're our wildest bequest to all the generations to come, and what is most to our credit is that the Wilderness Act is all about "ordinary people exercising grass-roots democracy."[548]

It does not leave the decisions of what lands will be protected to agencies like the U.S. Forest Service, National Park Service, or BLM.[549] Rather, wilderness protection decisions are made by Congress itself, usually only after enough

545 Backpacker magazine. "Extremes." *Backpacker*: September 2008, p. 116.
546 Duluth News Tribune. "Information about federal wilderness areas." *Duluth News Tribune*: 3/30/04.
547 Sierra Club. "Last Words." *Sierra*: July/August 2004, p. 72.
548 Doug Scott. "America has lobbyist to thank for wilderness." *Desert Morning News*: 2/26/06.
549 Doug Scott. "America has lobbyist to thank for wilderness." *Desert Morning News*: 2/26/06.

citizens stand up and say "enough" to the desecration and development of our rapidly dwindling wild public lands birthright. One of the underlying principles of the Wilderness Act is the democratic tenet allowing citizens to develop their own wilderness proposals and submit them directly to Congress.[550]

Oftentimes the impetus behind Congress adding another wilderness area to the Wilderness Preservation System comes from a single person looking to save a wild place they love. One Montana wilderness area was protected in 1972 thanks to the leadership of a small town hardware dealer; another in Nevada was preserved in 2002 due to the efforts of a casino cocktail waitress; and there are many more such instances, examples of just one person making the difference.[551]

With this in mind, Doug Scott says we should pause for a moment to appreciate the good we can all do in helping to lead a nation toward its highest aspirations: toward protecting its wilderness heritage by exercising our precious citizen democracy.[552] We follow (explains *Conservation Northwest* contributor Tim Coleman) in the steps of many wilderness visionaries: John Muir, Aldo Leopold, Howard Zahniser, Mardy Murie, and others. Like the Founding Fathers of our country, they wanted future generations to inherit the freedoms they had known, so they strove for decades to protect wild country, a path that we continue to tread.[553]

But you must, John H. Holcomb adds, "get involved to have an impact. No one is impressed by the won-lost record of the referee." Collective efforts create exponential results. As Wallace Stegner (the man Edward Abbey once pronounced, "The only living American writer worthy of the Nobel") said, "Talent lies around us like kindling waiting for a match, but some people, just as gifted as others, are less lucky. Fate never drops a match on them. The times are wrong, or their health is poor, or their energy low, or their obligations too many. Something ..." Don't let "something" stop you from making a difference.

In the great spirit and tradition of participative democracy, the Wilderness Act was the culmination of decades of collaboration among concerned citizens, wildlands conservation groups, environmentalists, hunters, anglers, and Congress—a diverse coalition of people and groups all looking to make a difference—and it established a new and long overdue standard for public lands protection. But wilderness is much more than just land; it is where our culture began, and as Wallace Stegner said, "We need wilderness preserved—as much of it as is still left,

---

550 Rick Hartman. "#22 The Wilderness Act." *Apex to Zenith*: 2nd Quarter 2008, p. 24.

551 Doug Scott. "America has lobbyist to thank for wilderness." *Desert Morning News*: 2/26/06.

552 Doug Scott. "America has lobbyist to thank for wilderness." *Desert Morning News*: 2/26/06.

553 Tim Coleman. "Achieving a lasting Columbia Highlands Wilderness." *Conservation Northwest*: Summer 2010, p. 5.

and as many kinds—because it was the challenge against which our character as a people formed."

Doug Scott adds: "The 'right' answer to how much wilderness 'should' be preserved is unknowable, but I will venture my own prediction: however much wilderness we Americans choose to designate and protect using the Wilderness Act, future generations are likely to judge not that we preserved too much, but that we preserved too little," and the ultimate challenge today is to save what's left, both here and abroad.[554]

Like the inspiration for national parks, wilderness preservation is an American construct, and today, this far-sighted idea has spread far from its American roots. Australia, New Zealand, Canada, Finland, Sri Lanka, the former Soviet Union, and South Africa, for example, have all passed similar legislation. Although not enacting wilderness protection legislation specifically, Italy, Zimbabwe, Namibia, and the Philippines have declared wilderness zones in public parks, municipal watersheds, game reserves, and forests.[555]

In 2004, the cabinet of Namibia approved a new national park comprising a major portion of the Sperrgebiet, a diamond mining reserve in southwestern Namibia. According to local news coverage, the government report said, "the Sperrgebiet fitted the definition of a wilderness 'as an area where the earth and its community of life are untrammeled by man, where man himself is a visitor who does not remain.'" As Doug Scott wrote in *The Enduring Wilderness*, "Howard Zahniser's words, adopted as the policy of the United States four decades ago, echo now around the globe."[556]

Despite originating the modern idea for national parks *and* wilderness areas here in the United States, wilderness still comprises only a small percentage of the country (as previously stated)—less than 3 percent of the Lower 48 states. Is that really enough? Is it enough room for all the other species that inhabit this great nation and continent? The country needs oil, coal, and lumber to supply industry for now, but I would argue that what remains of our wild, roadless public lands are more valuable to the nation left as they are.

Public opinion polling confirms strong public support for wilderness protection, regardless of region or party lines, and across the urban/rural divide.[557] More than six in 10 Americans believe there is not enough wilderness permanently protected for future generations. Furthermore, according to January 2003 polling conducted by Zogby International, more than two-thirds of respondents believe 10 percent or more of all lands in the United States should be protected

554 Doug Scott. *The Enduring Wilderness*. Golden, Colorado: Fulcrum Publishing, 2004, p. 124.

555 Joshua Tree National Park Guide: Spring 2004, p. 3.

556 Doug Scott. *The Enduring Wilderness*. Golden, Colorado: Fulcrum Publishing, 2004, p. 144.

557 Doug Scott. *The Enduring Wilderness*. Golden, Colorado: Fulcrum Publishing, 2004, p. 7.

as wilderness. When told that only 4.7 percent of the U.S. landbase has been permanently protected, nearly two-thirds feel it's "not enough."[558]

But wilderness has always been difficult to protect and designate, and it isn't getting any easier. As of 2004, when the Wilderness Act turned 40, opposition from motorized wreckreationists and energy development interests (the Blue Ribbon Coalition/wise abuse crowd) was stronger than ever, and the Bush administration had adopted many policies and passed some laws (while pushing dozens more) that undermined the wilderness system. Yet, as explained by Jeff Widen of the Colorado Environmental Coalition, "We still manage to move legislation."[559]

Why? Because those citizens "who see beyond our immediate, material wants and desires understand that wilderness is the very fabric upon which human history is written," and they refuse to give up, like Howard Zahniser, Wallace Stegner, Teddy Roosevelt, and tens of thousands of others who have followed in their footsteps.[560] Early in the twentieth century, President Theodore Roosevelt protected places like the Grand Canyon over the objections of short-sighted local politicians and developers. He refused to give up or give in, and he knew they would thank him some day. Teddy was right.

Today, Grand Canyon National Park and thousands of other national parks, monuments, wildlife refuges, and wilderness areas spread out across the United States and the world are tremendous environmental, recreational, and economic success stories. In March 1904, Forest Service Director Gifford Pinchot presented President Roosevelt with a report showing that the western states and territories with the most public land were "progressing rapidly in population and wealth." In other words, the larger the forest reserves, the more prosperity for a state or territory.[561]

In Seward, Alaska, residents originally condemned the creation of Kenai Fjords National Monument (established in 1978), which became a national park in 1980, but after reaping the financial and other rewards of living near protected public lands, they went back and said they wanted to expand the park.[562] The Seward Chamber of Commerce proudly proclaims this coastal town 125 miles south of Anchorage as "The Gateway to Kenai Fjords National Park."[563]

Henry Gannett, the former chief geographer of the United States Geological Survey (USGS), explored Kenai Fjords in 1899 and had this to say about the true

558 Friends of the Boundary Waters Wilderness (FBWW). "Preserving The Canoe Country Heritage." *FBWW*: May 2003.

559 Jeff Widen. "Wilderness Act celebrates 40 years." *The Colorado Environmental Report*: Summer 2004, p. 10.

560 Jeff Widen. "Wilderness Act celebrates 40 years." *The Colorado Environmental Report*: Summer 2004, p. 10.

561 Douglas Brinkley. *The Wilderness Warrior: Theodore Roosevelt And The Crusade For America*. New York: HarperCollins Publishers, 2009, p. 566.

562 Tracy Ross. "Ken Hits The Trail." *Backpacker*: December 2005, p. 20.

563 Bill Sherwonit. "A Warming Trend: After a Chilly Reception." *National Parks*: Winter 2005, p. 31.

"value" of such places: "There are glaciers, mountains and fjords elsewhere, but nowhere on earth is there such abundance and magnificence of mountains, fjords and glacial scenery... Its grandeur is more valuable than the gold or the fish or the timber for it never will be exhausted."

Clearly, in the long run, preserving nature is more profitable than exploiting it, as Representative Morris (Mo) Udall confirmed in 1988: "I've been through legislation creating a dozen national parks, and there's always the same pattern. When you first propose a park, and you visit the area and present the case to the local people, they threaten to hang you. You go back in five years and they think it's the greatest thing that ever happened. You go back in twenty years and they'll probably name a mountain after you."

As explained by *National Parks* contributor Thomas C. Kierman, in "The Value of Parks," the first national park established east of the Mississippi, Acadia, is on the rugged coast of Maine. The park's terrain covers just over 47,000 acres of granite-domed mountains, woodlands, lakes, ponds, and ocean shoreline, but per acre, a small national park like Acadia makes more of an economic impact ($3,400) than any acre of Maine's commercial forests ($368).[564] When Congress created Montana's Glacier National Park in 1910, some locals were hostile to the idea, fearing a park would "lock up" natural resources.

The Kalispell Chamber of Commerce and homesteaders opposed the park, saying it locked up timber and potential farmland that had "no particular scenic value" and would never attract tourists. Since then, Glacier has become the economic engine of northwestern Montana, now the most economically vibrant part of the state. The same results can undoubtedly be shown for national parks, monuments, wildlife refuges, roadless areas, and wilderness areas across the country and world.

In 1995, U.S. Forest Service economists measured the agency's holdings and found that national forests generated $125 billion a year in economic activity. An astounding 75 percent of that figure was based on recreation. Timber and mining, by contrast, amounted to 15 percent. By the 1990s, it had become all too apparent that recreation, not logging and mining, was the Forest Service's main product.[565] Don Barger, a senior regional director for the National Parks Conservation Association, noted that his organization did a study in 2006 and found that national parks nationwide contribute $4 to local economies for every $1 of federal money invested in them.[566]

In "As a Matter of Fact, Money Does Grow on Trees," *Outside* contributor

---

564 Thomas C. Kierman. "The Value of Parks." *National Parks*: May/June 2003, p. 6.
565 Chris Case. "What's Wilderness $$ Worth? Maybe Money Does Grow on Trees." *Trail & Timberline*: Spring 2010, p. 35.
566 Pam Sohn. "Public parks pay." *Chattanooga Times Free Press*: 3/28/10.

Bruce Barcott says no case study is more persuasive than Kane County, Utah, when considering the positive impact of protected public lands on local economies. Kane County is located about 210 miles south of Salt Lake City and is home to the Grand Staircase-Escalante National Monument, which was the site of one of Utah's most intense public lands battles. It started in 1996, when President Bill Clinton's creation of the 1.7 million-acre monument made him about "as popular in Kane County as a Yankee fan in Fenway."[567]

A Dutch-owned mining company was considering opening a coal mine on the monument's remote, grassy Kaiparowits Plateau, and locals expected the mine to boost the economy. When the monument designation scotched that plan, predictions of economic doom rang through the county seat of Kanab, population 4,500. Nearly 10 years later, writer-reporter Ray Rasker went to Kane County to see if the monument had crippled the local economy. It hadn't. In fact, Kane County was thriving.[568]

Rasker compared data from the four years prior to the monument's creation (1992-1996) with data from the four years after. During the latter period, the unemployment rate in Kane County dropped by more than half while labor income rose faster than it had in the premonument period. Per-job earnings, which fell 7 percent before the monument, rose 13 percent after it was created. Property values rose significantly too. "People in Kane County worried that all they'd get were low-wage tourism jobs," Rasker said. "In fact, the average wage per job went up."[569]

What accounted for the turnaround? "Word got out," says Rasker. "People read about Grand Staircase-Escalante. Some started visiting; others moved their businesses to Kanab or decided to retire in Kane County."[570] As explained by *Trail & Timberline* editor Chris Case (in "What's Wilderness $$ Worth?"), with the fortunes of the county hitched to the wellbeing of the mining industry and, therefore, at the mercy of globally determined price trends, Kane County *could* (and likely would) have suffered a roller-coaster economic ride. But now, best of all, the landscape that drew all of this attention is never going to pack up and leave the area.[571]

The Kane County experience is irrefutable proof that protected public lands are the ultimate sustainable economic engines, and Grand Staircase-Escalante isn't an isolated instance, here or overseas. Nations like New Zealand have shown that preserving the environment and setting aside national parks and

---

567 Bruce Barcott. "As a Matter of Fact, Money Does Grow on Trees." *Outside*: March 2005.
568 Bruce Barcott. "As a Matter of Fact, Money Does Grow on Trees." *Outside*: March 2005.
569 Bruce Barcott. "As a Matter of Fact, Money Does Grow on Trees." *Outside*: March 2005.
570 Bruce Barcott. "As a Matter of Fact, Money Does Grow on Trees." *Outside*: March 2005.
571 Chris Case. "What's Wilderness $$ Worth? Maybe Money Does Grow on Trees." *Trail & Timberline*: Spring 2010.

other protected public lands is a vital catalyst for sustainable economic growth, and having a "clean, green" image is attracting foreign tourists by the plane- and boatload.

In 2003, New Zealand's foreign tourism earned the country $4.4 billion, nearly 10 percent of their gross domestic product, and tourism provided one in 11 jobs, supported more than 15,000 businesses, and was the country's num- ber-one industry. Since 1992, international visitors have doubled.[572] Minnesota conservationist Paul Gruchow summed up the economics of wilderness preser- vation as eloquently as anyone: "The howl of a wolf, the cry of a loon, the lap of clean water against an untrammeled shore constitute the only common currency; to defend them is to labor in the most elementary way for the general good."[573]

There are similar economic benefits to be realized by countries weaning themselves off of fossil fuels. Robert F. Kennedy, Jr., lays out the facts. "We know that every nation that decarbonizes experiences instantaneous prosperity. For example, Iceland in 1970 was the poorest country in Europe. It was 100 percent dependent on imported coal and oil. Within 15 years, the government made Iceland completely energy independent, with 90 percent of their energy coming from geothermal sources. During that period, Iceland went from the poorest country in Europe to the fourth richest country by GDP on Earth."[574]

Even China has seen the light, as explained by *USA Today* contributor Calum MacLeod (in "China designates first of planned national parks"). "Snow-frosted trees, subzero temperatures, precious few people and one-story wooden houses. It sure looks like Siberia," says MacLeod, and it was for a generation of Chinese exiled there to labor in logging camps during the late 1960s. This land, close to the Russian border, where vast forests were stripped bare to help fuel the com- munist revolution, was designated as China's first national park in 2008.[575]

The move was inspired in large part by the USA, home to Yellowstone—the world's oldest national park. China's central government is now planning a parks system to curb destruction of the country's most scenic and biodiverse ar- eas. The Kanas Geological Park in Xinjiang province is fighting hard to become the next Chinese national park: 2.5 million acres of dramatic forests, mountains, and lakes, locally called "God's backyard."[576]

"Our park is the best in China, and will be even bigger than Yellowstone," says its head administrator Tan Weiping, who applied to Beijing for national park status. "We welcome U.S. tourists to come here," he adds. "And they can bring

572 Ray Lilly. "New Zealand." *The Desert News*: 4/17/04.

573 Friends of the Boundary Waters Wilderness (FBWW). "Preserving The Canoe Country Heritage." *FBWW*: May 2003.

574 Robert F. Kennedy, Jr. "Talking With...Robert F. Kennedy." *Nature's Voice*: March/April 2010, p. 4.

575 Calum MacLeod. "China designates first of planned national parks." *USA Today*: 11/21/08, p. 13A.

576 Calum MacLeod. "China designates first of planned national parks." *USA Today*: 11/21/08, p. 13A.

their tents, too, although we have star-rated hotels."[577] Zion National Park in Utah has since established a sister relationship with a national park in China.

Zion Superintendent Jock Whitworth said the agreement with Danxiashan National Park involves sharing ideas, staff, and research to better understand both parks. Danxiashan's director, Yu Changyong, said his park contacted Zion about a possible relationship because the two locations share a similar formation of red rocks. "Among all the red rock in America, Zion is the most diverse," he said. "We would like to work together for conservation and research of the red rock."[578]

Although the Chinese now understand the long-lasting economic benefits of protecting public lands, the Bush administration never did, or (more likely) simply didn't care. The positive economics of saving nature and creation is something they conveniently overlooked while busily shredding decades of environmental laws and conservation initiatives here in the U.S., often justifying their actions using an economic strategy called cost-benefit analysis. In 2002, the Bush administration said for the purpose of cost-benefit studies it would value each human life at $3.7 million dollars. Just two years earlier, the Clinton administration determined that a human life is worth $6.1 million—a figure that's 65 percent higher.[579]

It's especially ironic that the conservative "pro-life" Bushies would choose a formula that dramatically devalues human life—at least when it come to regulating the polluting and poisoning industries contributing to their campaign coffers.[580] In 2002, the Bush EPA also decided that the value of elderly people was 38 percent less than that of people under 70. After the move became public, the agency reversed itself.[581]

Was there any slimy level to which W wouldn't stoop in order to line the pockets of big industry? It seemed that just when we thought Bush couldn't possibly get any more vile and depressing, he pushed the envelope out once again. It was enough to make even a gonzo journalist like Hunter S. Thompson glum. "Bush depressed him; politics depressed him. He told me how embarrassed he was that his generation left this for mine," explains *Outside* contributor Emily Larocque.[582]

---

577 Calum MacLeod. "China designates first of planned national parks." *USA Today*: 11/21/08, p. 13A.
578 Associated Press (AP). "Zion forms relationship with nat'l park in China." *AP*: 12/14/09.
579 Osha Gray. "Nickel and Dimed." *Onearth*: Spring 2004, p. 38.
580 Osha Gray. "Nickel and Dimed." *Onearth*: Spring 2004, p. 38.
581 Seth Borenstein. "EPA's pronouncement: Your life now worth less." *Associated Press*: 7/11/08.
582 Emily Larocque. "Remembering Gonzo." *Outside*: November 2006, p. 38.

"They [the Bush administration] are the racists and hate mongers
among us–they are the Ku Klux Klan."
–Hunter S. Thompson[583]

From a regulatory and political perspective, devaluing human (and all other life) and undervaluing (or ignoring completely) the economic benefits of wild-lands preservation made it easier for W to justify more giveaways of our public lands, more pollution and pesticides, more development, and less regulation of industry. Author (of *The Grizzly Years*) and Vietnam veteran Doug Peacock said Bush was at war with all of the living, breathing world: "I think it's fair to say that there's never been an administration that's waged war against life more effec-tively than the Bush administration–human life, animal life, and wilderness."[584]

But we didn't take it lying down, and today, anybody and everybody can help fight back and make a difference. You, me, everyone can join with those of us who care about the future of humanity and life in general and help save what remains of the world's wild places and wide-open spaces. Just visit one or all of the sites below, sign up for their free e-mail action alerts (most have them), and use their *Take Action* centers to contact your elected legislators with the effortless click of a mouse.

The Wilderness Society–*www.wilderness.org*; The National Parks Conservation Association–*www.npca.org*; The Natural Resources Defense Council–*www.nrdc.org*; The Trust for Public Land–*www.tpl.org*; Trout Unlimited–*www.tu.org*; The Nature Conservancy–*www.tnc.org*; The Theodore Roosevelt Conservation Partnership–*www.trcp.org*; Backcountry Hunters and Anglers–*www.backcountryhunters.org*; REP America–*www.rep.org*; The League of Conservation Voters–*www.lcv.org*.

Doing this takes, at most, a matter of minutes and puts your elected legisla-tors on notice that you vote and are watching them. Just by doing this, you *will* be helping to make a difference in a world that's in desperate need of people who care about its future. REP member Bob Sanner cares. "Can we stop man's persistent trashing of the planet?" he asks. "Must we try? Of course! There is nobility in doing what is right in the face of futility. I chose nobility. May God bless everyone who makes a contribution to protecting our planet."[585] Wisconsin Senator Gaylord Nelson certainly made a contribution.

Senator Nelson showed the country and world the power of a single person with a simple idea, and he proved that a good one can change the world. In this instance, it changed because of Earth Day. *American Heritage* magazine <u>called the first</u> Earth Day, which drew 20 million participants, "One of the most

583 John Rogers. "Politically incorrect Thompson called an original." *Desertnews.com*: 2/22/05.
584 Tim Sprinkle. "Peacock and Bull." *Grist* magazine: 9/13/05.
585 Bob Sanner. "Write to REP..." *The Green Elephant*: Summer 2004, p. 12.

remarkable happenings in the history of democracy." Partly as a result, over the next 10 years, 28 major pieces of environmental legislation became law.[586] Not a bad legacy, Gaylord. Now it's up to us to follow in your footsteps, to protect and perpetuate your legacy, and to leave one of our own. But we have some big shoes to fill, as explained by *High Country News* contributor Tim Lydon.

"Nelson was elected senator in 1962, following 10 years as a state legislator and four years as Wisconsin's governor. An environmentalist before the term even existed, his first speech in Congress proposed banning phosphates in detergents that were choking lakes and rivers with rafts of foam. He also co-sponsored the Wilderness Act and proposed banning the carcinogenic pesticide DDT. His was an era of rising environmental awareness. Rachel Carson's 1962 book, *Silent Spring*, educated millions about how pesticides were killing the nation's birds.[587]

"In 1964, the Wilderness Act overwhelmingly passed Congress to become the strongest protection for federal lands. But it was also an era of startling environmental degradation. Pollution from cars and industry led to fatal smog events, including a 1965 episode that killed 80 New Yorkers. In January 1969, an oil spill near Santa Barbara blackened 30 miles of California's coastline. Six months later, Ohio's Cuyahoga River burst into flames, one of many 'river fires' caused by oil and chemical pollution..."[588]

"As Nelson pressed his agenda, the Earth Day movement grew, and on April 22, 1970, over 20 million Americans—fully 10 percent of the population—participated in demonstrations and teaching events. Rallies occurred in scores of cities, over 12,000 schools held events and Congress recessed in honor of the day. The whole thing was as American as apple pie..."[589]

"In the wake of Earth Day, leaded gasoline, DDT and other pollutants were banned. Congress passed landmark environmental laws still essential today, including the 1972 Marine Mammal Protection Act, 1973 Endangered Species Act, 1974 Safe Drinking Water Act, and others. By the time Nelson left Congress in 1981, key parts of his environmental agenda had become law. Afterward, Nelson became a counselor for The Wilderness Society, and in 1995 he was awarded the Presidential Medal of Freedom."[590]

---

586 The Wilderness Society. "A Fond Farewell to the Father of Earth Day." *Wilderness*: 2005-2006, p. 59.
587 Tim Lydon. "Only 40 years ago, the Earth got its day." *High Country News*: 4/13/10.
588 Tim Lydon. "Only 40 years ago, the Earth got its day." *High Country News*: 4/13/10.
589 Tim Lydon. "Only 40 years ago, the Earth got its day." *High Country News*: 4/13/10.
590 Tim Lydon. "Only 40 years ago, the Earth got its day." *High Country News*: 4/13/10.

Former U.S. Supreme Court Justice William O. Douglas was another self-less campaigner for the wild and natural world. "When the Earth, its products, its creatures become his concern," Douglas said, "man is caught up in a cause greater than his own life and more meaningful. Only when man loses himself in an endeavor of that magnitude does he walk and live with humility and reverence." Today, although no longer with us, Senator Nelson and Justice Douglas would surely say, "Whatever you do, don't just sit back and do nothing. Instead, make a difference by fighting against the apathy and indifference all too prevalent in our country and world."

> "The key is endless pressure, endlessly applied. Flow
> like the water, and when you hit a barrier, flow
> around it. Keep on flowing. Keep on going.
> The river is going to find a path."
> –Brock Evans[591]

AUGUST

During the first weekend of August, I found myself on a path to the summits of Mounts Belford and Oxford, hiking-climbing both peaks on August 5. Mount Belford was named after James Burns Belford, a Colorado territorial judge appointed by President Ulysses S. Grant. Mr. Belford was also Colorado's first member of the U.S. House of Representatives. In Washington, he was nicknamed "the Red-Headed Rooster of the Rockies" because of his colorful manner and appearance.[592] As for Mount Oxford's name, I'll let you make an educated guess.

Belford and Oxford are two of Colorado's least difficult 14ers, but they're still immense piles of rock over 14,000 feet high, and even though I subscribe to Reinhold Messner's fast and light climbing mantra, no speed records were broken or paradigms shaken on this day. In fact, I had a difficult time completing this *twofer* for some reason, but the best things in life rarely come easy, and no 14ers are effortless hikes or climbs. Mounts Belford and Oxford are situated such that on the standard route, you climb Belford to get to Oxford.

From Belford, drop down to the connecting saddle between the two peaks, ascend Oxford, and then repeat in reverse. Starting from the Missouri Gulch trailhead, the route is Class 1 rated, but transitions to Class 2 on Belford's west slope. Many climbers choose to do Mount Oxford after Belford because the two

---

591 Jamie Sayen. "Interview with an Activist: Brock Evans." *North Woods Vision*: Spring 2000, p. 17.
592 William Bright. *Colorado Place Names* (3rd Edition). National Geodetic Survey.

mountains are only about a mile apart, but consider the local weather conditions before making a commitment to Oxford. You'll likely be on the connecting saddle for a few hours, and if a storm moves in, there's no good place to seek shelter.

Missouri Mountain is also close by, but if you plan to do all three peaks in a single day, be prepared for a good workout and a lot more exposure to the natural forces that frequently manifest themselves when "rock meets sky." Doing this *threesome* is a 14½-mile jaunt with a combined 7,400-foot elevation gain. The Belford-Oxford combination is a more modest 11 miles through 5,900 feet of elevation gain, and Belford alone is seven miles with a gain of 4,560 feet.[593]

It takes about 1½ hours to reach Oxford from Belford, and generally a bit longer coming back. As the folks at Summitpost.org warn, this round-trip has you wholly exposed on a high mountain ridge for a long time, so try getting an early morning start to reduce your chances of becoming a lightning rod. Expect to spend anywhere from nine to 11 hours in the mountains to complete the combo. This is a long day, and a predawn start will increase your chances of success and safety.[594]

Bruce Morrow, education and outreach manager for the Colorado Fourteeners Initiative, offers these useful tips for climbing fourteeners: wear appropriate footwear—waterproof or water-repellant boots with full-leather uppers are best. This isn't just a suggestion to keep your feet more comfortable. Hikers wearing tennis shoes or hiking boots that are part fabric tend to skirt muddy or snow-covered patches of trail so their feet won't get wet. In the process, they create new and unnecessary trails on the fragile tundra.[595]

Sometimes, Morrow adds, the results can be deadly, to both people and the environment. "There was an incident on Huron Peak, where a woman with inappropriate footwear sprained her ankle," he explains. "Someone called for a rescue from a cell phone, and by the time the communication got to the sheriff it sounded very serious." A helicopter was dispatched and crashed, killing the pilot and flight nurse. "Aside from a lot of people being affected personally, there was aviation fuel and debris all over the side of Huron," Morrow said.[596]

Rest assured, I have good, waterproof footwear, but I still found the 11-mile Belford-Oxford combo far more exhausting than it should have been given my fitness level, and I barely mustered the energy to reascend Belford on the return

---

593 http://www.summitpost.org/show/mountain_link.pl/mountain_id/344.

594 http://www.summitpost.org/show/mountain_link.pl/mountain_id/344.

595 Deb Accord. "For The Love Of...Fourteeners/Climbers who scale 54 tallest peaks." *The* [Colorado Springs] *Gazette*: 4/19/01.

596 Deb Accord. "For The Love Of...Fourteeners/Climbers who scale 54 tallest peaks." *The* [Colorado Springs] *Gazette*: 4/19/01.

trip. As singer Mary Chapin Carpenter sang, "Some days you're the windshield, some days you're the bug." I felt like the bug hitting a windshield at high speed climbing Belford on the return trip, but gritted my teeth and dug deep down for the reserves of strength and stamina most of us can muster when truly motivated to complete a difficult and rewarding task.

Over the years, I've hiked and climbed with a handful of people who are unable to do this; they can't or won't push themselves when things get tough; they quit mentally long before it's physically necessary. Needless to say, I don't climb with them anymore. According to *Backpacker* contributor Andrea Lankford, National Park Service rangers call hikers like this a "Code W." "A Code W is a wimp. There is nothing medically wrong with a Code W. His spirit, not his body, is broken," she explains.[597]

Both physical and mental toughness, combined with all-around focused discipline, are necessary to be a successful climber or mountaineer (not to mention successfully negotiating life in general), and those who can't or won't hack it are a danger to themselves and their hiking-climbing companions. As *National Geographic Adventure* contributor Rick Newton wrote, "A life without struggle is a life without lessons, without learning."[598]

Ansel Adams was a lifelong learner and once said, "The intense spirit back of all mountain adventure, the toil, the keen delights, the good-fellowship, the humor—the underlying golden humanity that reveals itself in the presence of hardship, repulse, and death—one wonders if there is any human activity that tests strength of body and soul more severely than man's struggle with the stern crags and glaciers, or that yields such true satisfaction when his intense efforts result in success."[599]

I returned to the Missouri Gulch trailhead with my 18th and 19th fourteeners in the bag, and then headed for Mount Harvard. Mount Harvard (14,420 ft.) was named in honor of Harvard University by alumni who were members of an 1869 survey party. Other nearby peaks were also named by graduates of eastern schools, and the grouping eventually became known as the Collegiate Peaks.[600]

Mount Princeton, the southernmost of the Collegiates, measures one foot higher than neighboring Mount Yale, and, thus, is reported to be the subject of summit rock removal to the tune of two feet by certain Ivy Leaguers.[601] According to Summitpost. org, Mount Harvard was named when Harvard geology professor Josiah Whitney

---

597 Andrea Lankford. "Ranger Confidential." *Backpacker*: September 2010, p. 88.
598 Rick Newton. "Life." *National Geographic Adventure*: April 2002, p. 13.
599 Ansel Adams. *Sierra Club Bulletin*, 1928.
600 David Covill and John D. Mitchler. *Hiking Colorado's Summits.* Guilford, Connecticut: The Globe Pequot Press, 2002, p. 91.
601 Randy Jacobs (ed.). *Guide to the Colorado Mountains.* Golden, Colorado: The Colorado Mountain Club Press, 2000, p. 195.

led a surveying expedition into Colorado's mountains to investigate rumors of soaring 17,000-foot peaks deep in the Rockies. After crossing Trout Creek Pass, they named the highest summit in sight for the expedition's sponsor.[602]

Mount Harvard is the third-highest peak in Colorado and only one of three that rise above 14,400 feet. It's also the fourth-highest peak in the contiguous 48 states and the highpoint of Chaffee County. Harvard is located 11 miles northwest of Buena Vista, not readily visible from encroaching roads or highways, and is often climbed in combination with its neighbor, Mount Columbia. The combo is not a slam dunk though, because the ridge between these Collegiate Peaks is 2.2 miles long with significant elevation gain and loss.[603]

Virginia Simmons (author of *The Ute Indians of Utah, Colorado, and New Mexico*) says she admires the Collegiate's beauty, but "their names smack of—well—arrogance. The dust of Indian ponies had scarcely settled before Ivy Leaguers were scampering up and down the mountains in heady enthusiasm, bestowing the names of their alma maters on them. We can be grateful that the Utes at least received token recognition in the Indian Group (Shavano, Tabeguache, and Antero)."[604]

There also don't seem to be many Colorado peaks named after women. A proposal to name an unnamed Sawatch Range thirteener was put before the U.S. Board on Geographic Names during October 2006. The summit lies on a ridge that separates fourteeners Mount Antero and Tabeguache Peak. The commemorative name would honor one of Colorado's first female mountaineers, and a charter member of the Colorado Mountain Club: Agnes Vaille.[605]

The suggested name for this 13,591-foot peak is "Point Agnes Vaille." Vaille (1890-1925) was a well-known mountain climber who pioneered numerous routes in the Rocky Mountains. She died during a winter blizzard after climbing the east face of Longs Peak in January 1925. Her death (as previously mentioned) is memorialized on a plaque at the Keyhole Shelter (13,150 ft.) on the northwest ridge of Longs Peak.[606] Here *Denver Post* writer Steve Lipsher provides a 2007 update on the Point Agnes Vaille proposal.

"An effort to name a Colorado high peak after a pioneering female mountaineer has proved to be an uphill battle for proponents. On the surface, the request by the Colorado Mountain Club is simple: Dedicate

---

602 http://www.summitpost.org/mountain/rock/150370/mount-harvard.html.

603 http://www.summitpost.org/mountain/rock/150370/mount-harvard.html.

604 Virginia McConnell Simmons. "Naming the Indian group of the Sawatch Range." *Colorado Central Magazine*: June 2005.

605 http://www.fourteenerworld.com/.

606 http://www.fourteenerworld.com/.

an unnamed 13,591-foot Chaffee County mountain for Agnes Vaille, a celebrated alpinist who perished on a risky winter ascent of Longs Peak in 1925. But at a time when the number of unnamed summits in Colorado is dwindling, the proposal has met resistance from Chaffee County commissioners and the state Board of Geographic Names.[607]

"The county commissioners point out Vaille already has a scenic waterfall and a stone shelter on Longs Peak named in her memory. 'Let's save some geologic features to bear the names of other past and future Coloradans,' wrote a commentator named 'Iceman' on the 14ers.com online forum. The final decision will be made by the U.S. Geological Survey's Board on Geographic Names, which tends to defer to local wishes and historic or popular designations.[608]

"The push to name peaks in Colorado is colliding with shrinking real estate. Only seven of the highest 100 peaks in Colorado still do not have names. Would-be namers now are seeking less-prominent subsidiary peaks. There are 20 naming proposals pending for peaks in Colorado before the federal board..."[609]

"Vaille was renowned as a bold climber who was well on her way to becoming the first woman to scale all of the state's 14,000-foot peaks. Vaille, a secretary with the Denver Civic and Commercial Association, was 34 when she froze to death on the slopes of Longs Peak on Jan. 12, 1925... 13,870-foot Cronin Peak was named after Vaille's frequent climbing partner, Mary Cronin."[610]

David Covill and John Mitchler, authors of *Hiking Colorado's Summits*, note that more 14,000-foot peaks are located in Chaffee County than in any other place in North America. The county is home to 12 fourteeners, seven more than runner-up Hinsdale County, Colorado, with a mere five, and two more than larger Inyo County, California, with 10.[611] But to its lofty credit, Inyo County has the Lower 48's highest peak—14,495-foot Mount Whitney—amongst its fourteeners. California purportedly also has the last fourteener to be climbed in the Lower 48: Thunderbolt Peak (14,003 ft.) was first climbed on August 13, 1931.[612]

Chaffee County and other mountain-strewn locales gain more than bragging rights from having so many fourteeners in their midst. One Colorado study

607 Steve Lipsher. "Uphill battle to name Chaffee County peak." *Denver Post*: 6/9/07.
608 Steve Lipsher. "Uphill battle to name Chaffee County peak." *Denver Post*: 6/9/07.
609 Steve Lipsher. "Uphill battle to name Chaffee County peak." *Denver Post*: 6/9/07.
610 Steve Lipsher. "Uphill battle to name Chaffee County peak." *Denver Post*: 6/9/07.
611 David Covill and John D. Mitchler. *Hiking Colorado's Summits*. Guilford, Connecticut: The Globe Pequot Press, 2002, p. 91.
612 http://www.fourteenerworld.com/.

showed that the average person climbing a fourteener spends $200 within 25 miles of the summit during their trip. Given that Chaffee County encompasses 12 14,000-foot peaks, The Quiet Use Coalition (www.quietuse.org) calculates that 14er climbers add over $12 million to the local economy annually.[613]

Climbing Mount Harvard and injecting some cash into Chaffee County's coffers was next on my 14er agenda (on August 12), and Huron Peak (my 21st fourteener) followed on August 26. About climbing Harvard, Gerry Roach says: "Like most Sawatch fourteeners, Harvard is an easy climb, but requires more effort than most. Many parties choose to pack in when climbing Harvard. Harvard will both test and reward you."[614] That it did, and after summiting her, I set my sights on Huron Peak.

Although she barely qualifies as a fourteener (14,012 ft.), Huron Peak is the 14er farthest from a paved road in the Sawatch Range, an accolade that more than makes up for any perceived height-related shortcomings. Gerry Roach describes Huron as "a shapely, shy peak hidden in the heart of the Sawatch about halfway between Buena Vista and Independence Pass. Huron just barely rises above 14,000 feet but... is the farthest from a paved road. The view from Huron's summit is one of the best in the Sawatch."[615] The shortest and most popular Huron Peak route is on the South Winfield Trail, an eight-mile round-trip Class 2 hike-climb through an elevation gain of 3,740 feet.

I honestly don't remember anything particularly distinctive about climbing Huron, but I know it was beautiful because the views from high atop the summit of any 14er in the mountains of Colorado, or anywhere else, always are. And reaching the summit of any mountain anywhere is, without a doubt, a rewarding and momentous event—the icing on a climber's cake—but it's the cake, the experience as a whole, that's most filling.

Still, some may ask why these mostly solo wilderness hiking and climbing adventures are so special. Why all the vagabond wandering when you could be sitting at home with a loved one watching kids and television? My answer: while climbing, for short periods of time, I become the proverbial lone wolf, roaming wild and daunting landscapes, needing only minimal sustenance and road- and OHV-free scenery to be happy; shunning comfortable surrounds and embracing the indifferent but incredibly calming salve of wilderness. *That's why,* for starters.

Fourteeners completer Dave Philipps knows why. "Climbing the fourteeners can take hikers to remote spots in the state they might never otherwise see," he says. "Past hidden alpine lakes and roaring waterfalls, and up along sculpted

613 The Quiet Use Coalition. "Facts about Our Trails and Quiet." *Quiet Times:* Spring 2007, p. 3.
614 Gerry Roach. *Colorado's Fourteeners.* Golden, Colorado: Fulcrum Publishing, 1999, p. 118.
615 Gerry Roach. *Colorado's Fourteeners.* Golden, Colorado: Fulcrum Publishing, 1999, p. 103.

ridges with views stretching 100 miles. The struggle to the summit becomes addictive."[616] Ansel Adams was a mountain climbing addict too. "A huge mountain cannot be denied—it speaks in silence to the very core of your being."[617] It also beats the heck out of lethargic, dumbing-down, brain-cell-slowing television watching and most other sedentary pastimes. *That's* why.

In *Sandstone Sunsets*, Mark Taylor throws his opinion into the ring. "Pockets of profound silence can still be found... This silence breaks on the ear like the tide of a great ocean... I have seen city people so unnerved by it that they have packed up and left—never to return... others act as if they have finally arrived back at a place they once knew but lost so very long ago."[618] Joining the likes of Mark Taylor, John Muir, Doug Peacock, David Petersen, Chris McCandless, and Aron Ralston in their drive to experience wilderness alone and on its terms, I seek out transcendent solitude and leave the dragging anchor of humanity behind. *That's* why.

Mary Roberts Rinehart knew why too. "Cities call—I have heard them. But there is no voice in all the world so insistent to me as the wordless call of the Rockies. I shall go back. Those who go once always hope to go back. The lure of the great free spaces is in their blood."[619] She ends with, "Great spaces; cool, shadowy depths in which lie blue lakes; mountainsides threaded with white, where, from some hidden lake or glacier far above, the overflow falls a thousand feet or more, and over all the great silence of the Rockies."[620] *That's* why.

Some peak baggers are also surely working on life lists, but "I'm not the most compulsive list maker," says Steve (centennial completer) Mueller. Still, "it's a measurable way of saying you've done something different; you've been able to do something challenging where you can set a tough goal and achieve it," he adds.[621] Fourteener completer Carlton Stoiber puts another layer of icing on the "Why climb 14ers?" cake. "I recognize that these 54 peaks represent a somewhat artificial set of goals—but that could be said of many of life's aims."[622]

Beside (as Daniel Burnham said), it's best to "make no little plans; they have no magic to stir men's blood."[623] *That's* why. But what about ego, which is what one of my friends seems to think drives me to climb? Fourteener (and 13er/12er) completer Ken Nolan has an appropriate response: "Ego? That's

---

616 Dave Philipps. "Fourteeners 101." *The* [Colorado Springs] *Gazette*: 3/28/07.

617 Ansel Adams. *Sierra Club Bulletin*, February 1932.

618 Mark A. Taylor. *Sandstone Sunsets*. Salt Lake City: Gibbs-Smith Publishers, 1997.

619 David Harmon (ed.). *Mirror of America*. Boulder, CO: Roberts Rinehart Publishers, 1989, p. 51.

620 David Harmon (ed.). *Mirror of America*. Boulder, CO: Robert Rinehart Publishers, 1989, p. 49.

621 Lori Spaulding. "Centennials take climbers beyond 14ers to other challenging high points." *PikesPique*: September 2009, p. 1.

622 Linda K. Crockett. "the fourteener files." *Trail & Timberline*: Winter 2009, p. 17.

623 Tom Wohlgenant. "From the Director." *The Trust for Public Land-Colorado*: Summer 2008, p. 2.

obviously a factor, but the number of people who are impressed or even care is so small that recognition is rather elusive."[624]

Colorado climber Dave Haynes comes as close to hitting the nail on the head as anyone regarding this "Why climb" conundrum. Dave said (in *The [Grand Junction] Daily Sentinel*), "Once you let go of the need to have a reason for everything, trudging to the top of a tall peak seems like a good idea. If you're going to walk all day, why not be rewarded with an amazing view?"[625] In early September it was time to trudge to the top of Mount Lindsey (my 22nd fourteener), to another amazing view.

September

Mount Lindsey was named (in 1954) after Malcolm Lindsey, a devoted leader of the Colorado Mountain Club's Denver Juniors Group.[626] Overshadowed by Blanca Peak and Ellingwood Point, Malcolm's namesake rests by itself 2½ miles east of Blanca, with its summit and southern slopes on private property.[627] Mount Lindsey isn't one of the better known Colorado fourteeners, but it is surprisingly one of the more difficult peaks to climb: the easiest route open to the public is a very loose Class 2+ scramble.

The terrain surrounding Lindsey includes some of Colorado's trademark dramatic scenery, and the Huerfano Valley is worth a visit even if climbing isn't on your mind. From the valley floor, Blanca and Ellingwood peaks' shared northeast face dominates the skyline, and the aptly named 13,500-foot Iron Nipple dominates Lindsey's standard approach. Because part of Mount Lindsey is privately owned, the Huerfano Valley's North Face and Northwest Ridge are the only accessible climbing routes, and they occasionally take lives, as this October 2000 story from Erin Smith of *The Pueblo Chieftain* confirms.

> "The body of a 58-year-old hiker will have to remain where the man died on Lindsey Peak because it is too dangerous to attempt to remove... Costilla County Sheriff John Mestas said Tuesday. 'Unfortunately, we will not be able to get him out,' Mestas said. Saturday evening...David Lee Syring of Brighton fell to his death on his way down the peak and was partially buried under an avalanche of rock at about 13,200 feet... some

---

624 Chris Case. "Why the art (or is it obsession) of peak bagging." *Trail & Timberline*: Winter 2009, p. 21.

625 Dave Haynes. "Peakbagging's not for everyone, but to summit's a big deal." *The [Grand Junction] Daily Sentinel*: 9/13/09.

626 http://www.fourteenerworld.com/.

627 Gerry Roach. *Colorado's Fourteeners*. Golden, CO: Fulcrum Publishing, 1999, p. 180.

800 feet below the peak's summit.[628]

"Mestas said he has been in contact with Syring's son and son-in-law, and said the Syring family asked a private Durango rescue group to retrieve the body. But the private group, after going to the area, agreed with three search and rescue units from throughout the state that it was too dangerous to do so. Mestas said the body is in an 'avalanche chute on the edge of a cliff where there are spontaneous rock slides. It's too risky. We can't risk the health and safety of rescue workers.'"[629]

I don't know if David's body was ever recovered, but I'll vouch for the dangerous conditions on Mount Lindsey. I climbed her cautiously—a mere month before David Syring died there—and after completing this steep, scree-strewn route, I took a breather until October. Time to relax sore muscles and contemplate the meaning of life. Over the years, my mostly solitary pursuits have resulted in increasingly fewer opportunities for social interaction.

All of my married friends now have kids, not that I saw that much of them before, and have settled down into the seemingly monotonous (to me) routine of family life. So when I'm not in the mountains or traveling elsewhere on weekends, I walk over to a neighborhood bar, The Hatchcover, for four or five or more beers and some oversalted but oh-so-good popcorn.

Usually making minimal effort to interact with the locals, I quietly watch whatever football, hockey, baseball, or basketball game is on television—interesting for an hour or so only because I haven't had cable or satellite television for 20 years—while my mind wanders the woodlands and wildlands, the mountains and canyons, the trials and tribulations of my vagabond-wannabe life. When someone asks who's playing or what the score is, I reply truthfully: "I wasn't paying attention." After four-plus years of frequenting The Hatchcover, I've assembled a good group of beer drinking buddies and acquaintances, but I still prefer to get in and out quickly, before the place gets busy and bad karaoke singers take to the stage.

So much for my social life. A curse or a blessing, I'm not sure, but I wouldn't change it for the world. Like my friend David Petersen, "I've never been the sort, even when young and reckless, to care for the barroom scene, with its over-amped music and crowded dance floors; the stifling stench of cheap aftershave, industrial-strength perfume, and raging pheromones; cover charges, drunken boring patrons, and steroidal bully bouncers."[630] For me (like David, it seems), it's off-the-beaten-path, hole-in-the-wall-type bars or none at all.

628 Erin Smith. "Hiker's body may have to remain in mountains." *The Pueblo Chieftain*: 10/04/00.
629 Erin Smith. "Hiker's body may have to remain in mountains." *The Pueblo Chieftain*: 10/04/00.
630 David Petersen. *On The Wild Edge*. New York: Henry Holt and Company, 2005, p. 16.

Conversely, surely there's something to be said for those in the perpetual pursuit of relationships; for nonstop barhopping and pub crawling, compulsive dating and one-night stands; for dining out, watching TV, and going to movies, but I have more worthwhile things to do with my life for now, and hopefully forever, as long as that lasts for the likes of me. Despite having some good friends at the Hatchcover, I find most (but not all) of the locals uninteresting.

They are (for the most part) the voiceless (by choice) masses who sit idly by, apathetically watching the world circle the drain. Content with their cookie-cutter homes, cardboard condos, and overpriced entertainment centers with 100 channels of mind-numbing garbage, these are the people who are destroying the planet by default, because they won't lift a finger to help save what's left. Desensitized to the wider, wilder world around them, hooked on so-called reality and other TV shows, they can't comprehend that in reality, they're the ones most responsible for destroying the planet in slow motion.

Hunter, political activist, and Motor City Madman Ted Nugent questions the mentality of these people in his book *God, Guns & Rock'N'Roll*. "I cannot for the life of me understand the mindset that allows American citizens in this awesome experiment in SELF government to just go about a repetitious, day in and day out existence like so many sheep. That, to my thinking, is virtual insanity. The bad rut. Then you die. In reality, a practical death comes much sooner than a biological death, for to merely exist is in fact not to live at all."[631] Well said, Ted.

David Petersen adds his thoughts to the fray, saying, "We only live once, and most of us don't even do that."[632] These unthinking, uncaring human lemmings wrapped in thought-stifling cocoons of despondent ignorance, multiplied by millions, are decimating the planet. In lieu of socializing with them, it's easier and more enjoyable to drink my four or five beers in peace and quiet, in silent contemplation, and then go home and get a good night's rest in order to wake better prepared to face another day in a world filled with too many people just like them.

Ahhhh, the damn frustrating futility of it all. As R. L. Dabney said about similar people over one hundred years ago, "If you teach everyone to read, all you will accomplish is to create a mass market for trash literature." Dabney was Civil War General Stonewall Jackson's pessimistic but realistic chaplain.[633] My friend Minnesotan and Civil War buff Barry Babcock adds a twenty-first-century twist to Dabney's observation: "I read about... all these red-necks saying global warming is a hoax. It leads me to believe that hydrogen is not the most abundant element in the universe... it's human stupidity."[634]

631 Ted Nugent. *God, Guns & Rock'N'Roll*. Washington, D.C.: Regnery Publishing, Inc., 2001, p. 82.
632 David Petersen. *On The Wild Edge*. New York: Henry Holt and Company, 2005, p..xviii.
633 Charley Reese. "Words are powerful tools in the hands of a spinner." *The Vicksburg Post*: 8/27/06, p. A4.
634 Barry Babcock. "Re: Antarctic ice shelf set to collapse due to warming." *E-mail*: 1/24/09.

Why do I bother? How do I keep going? In part, because I like a challenge, even when facing seemingly insurmountable odds. I prefer a David vs. Goliath-like contest; I thrive on the near-futility of it, because as a Chinese proverb reminds us, "A gem cannot be polished without friction, nor man perfected without trials." In *A Pirate Looks at Fifty*, Jimmy Buffett puts the icing on this distasteful cake: "It's small minds and small thinkers who don't consider anything outside of their neighborhood important. We are all citizens of the world. You'd better come to terms with that concept or get ready to spend your life behind a fast-food counter."[635]

And Eleanor Roosevelt lights the candles: "Great minds discuss ideas; Average minds discuss events; Small minds discuss people." So be it. At least their indifferent laziness keeps them out of the mountains and far away from me when I choose to escape the groping clutches of the gone-astray civilization they're mindlessly embracing and perpetuating, and most of the mountain climbing I do is by no means technically challenging or overly difficult, nor does it require using ropes or having spiderlike agility. It just takes a firm resolve to get up off your butt and out from in front of the television, combined with a moderate amount of effort and determination to do more than waste (physically and mentally) away.

In fact, as of the autumn of 2000, I hadn't climbed any routes on peaks requiring the use of technical climbing aides. For me, climbing is more about the spectacular views than the bitter-sweet challenge; the "man against mountain" mentality. In fact, I'm not a climber in any true technical sense, but instead (at best), a middle-weight explorer-adventurer. I learn only what I need to in order to see what I want to see and do what I want to do, no more. John Muir, one of America's earliest climbers and a minimalist when it came to using climbing gear, would likely agree with my philosophy. Muir was a natural and fearless climber, as this description from his friend Reverend S. Hall Young confirms:

> "Then Muir began to *slide* up that mountain. I had been with mountain climbers before, but never one like him. A deer-lope over the smoother slopes, a sure instinct for the easiest way into a rocky fortress, an instant and unerring attack, a serpent-glide up the steep; eye, hand and foot all connected dynamically; with no appearance of weight to his body—as though he had Stockton's negative gravity machine strapped on his back.[636]
>
> "Fifteen years of enthusiastic study among the Sierras had given him the same preeminence over the ordinary climber as the Big Horn of the Rockies shows over the Cotswold. It was only for exerting myself to the

635 Jimmy Buffett. *A Pirate Looks at Fifty*. New York: Ballantine Publishing Group, 1998, p. 327.
636 Lee Stetson (ed.). *The Wild Muir*. Yosemite National Park, CA: 1994, p. 155.

limits of my strength that I was able to keep near him... And such climb-ing! There was never an instant when both feet and hands were not in play, and often elbows, knees, thighs, upper arms, and even chin must grip and hold.[637]

"Clambering up a steep slope, crawling under an overhanging rock, spreading out like a flying squirrel and edging along an inch-wide pro-jection while fingers clasped knobs above the head, bending about sharp angles, pulling up smooth rock-faces by sheer strength of arm and chin-ning over the edge, leaping fissures, sliding flat around a dangerous rock-breast, testing crumbly spurs before risking his weight, always going up, up, no hesitation, no pause—that was Muir!"[638]

In his essay "Rebel Yell: Vertical Urban Guerilla Warfare in the Deep South," (in *The Climbing Art*: #31, p. 89), Mac Hester describes one of his friend's climb-ing technique with equally lyric (but less lengthy) prose: "Then he launched his whippet-like body onto and into the crack and climbed with a smoothness and grace which cannot be adequately expressed in words but may be tasted in good bourbon."[639] Sounds like something (the good bourbon) I should try sometime.

OCTOBER

During early October, I ventured east, halfway across the country, to Mount Katahdin in Maine's Baxter State Park. The park's founder, Percival P. Baxter, was Maine's governor from 1921-1924. As a child, Percival enjoyed fishing and vacationing in the Maine woods, and his affection for the land and wildlife was instrumental in his protecting Baxter State Park for the people of Maine. He began to fulfill his dream in 1930 with the purchase of 6,000 acres, including Mount Katahdin, Maine's highest peak.[640]

In 1931, Baxter formally donated the parcel to the State of Maine, with the con-dition that it be kept "forever wild." Today, Baxter State Park encompasses 204,733 acres, and its wild core (150,564 acres) is managed as a wildlife sanctuary. In the northwest corner, a 28,594-acre parcel was designated by Governor Baxter to be managed as a Scientific Forest Management Area (SFMA): a showplace for sound forestry, hunting, and trapping. There are an additional 22,906 acres outside the SFMA where hunting and trapping are allowed under the park's deed.[641]

---

637 Lee Stetson (ed.). *The Wild Muir*. Yosemite National Park, CA: 1994, p. 157.
638 Lee Stetson (ed.). *The Wild Muir*. Yosemite National Park, CA: 1994, p. 157.
639 Mac Hester. "Rebel Yell: Vertical Urban Guerilla Warfare in the Deep South." *The Climbing Art*: #31, p. 89.
640 http://www.baxterstateparkauthority.com/aboutus/history.html.
641 http://www.baxterstateparkauthority.com/aboutus/history.html.

Mount Katahdin received its name, which means "The Greatest Mountain," from the Penobscot Indians.[642] In 1846, Henry David Thoreau scaled Mount Katahdin, located 25 miles north of present-day Millinocket, surveying from the top an "immeasurable forest for the sun to shine on. No clearing, no house. It did not look as if a solitary traveler had cut so much as a walking-stick there."[643]

Thoreau, it seems, was something of a peak bagger. Beginning in 1839 with Mount Washington in New Hampshire, he made some 20 ascents of New England mountains before 1860: Greylock and Wachusetts in Massachusetts; Lafayette, Monadnock, Red Hill, Temple, Uncanoonuc, and Wantiasiquet in New Hampshire; and Kineo in Maine. An initial Mount Katahdin attempt left him lost in pea-soup fog with a sprained ankle, but produced the lyrical essay *Ktaadn* (later published in *The Maine Woods*).[644]

While not as "great" in height as a 14er (actually, Mount Katahdin is only a 5er), she's just as difficult to climb as many of Colorado's fourteeners. *Backpacker* magazine cautions those who may contemplate Katahdin as an afternoon stroll: "This horseshoe-shaped peak is the northern terminus of the AT [Appalachian Trail], but also a great scramble in its own right: The infamous Knife Edge route has plenty of dazzling exposure. Check your fear of heights at the trailhead."[645]

But because it's less than half the height of high peaks in our Rocky Mountains, Katahdin is definitely a less physically demanding hike-climb. In fact, downtown Colorado Springs (at just over 6,000 ft.) is higher than Mount Katahdin (5,267 ft.). Still, not bad for the East Coast. Someone living in, say, New York City, would have to ride up an elevator the equivalent of nearly five Empire State Buildings just to reach the top of Mount Katahdin.

More significant is the fact that nothing in New York City or vicinity can hold a candle to the spectacular scenery and wildlands of Mount Katahdin and northern Maine. As *Backpacker* contributor Mike Magnuson said (in "Retro on the Ridge"), "Up here, I can see I-don't-know-how-far into the distance, rows and rows of mountains extending forever in such a way that makes the earth look like a tree-covered river flowing into eternity."[646] However, New York has some wild country too. By approval of an 1894 referendum, New Yorkers added the now-famous "forever wild" clause to their state constitution:[647]

642 D'Arcy Fallon. "Highpoint Hiking." *American Hiker:* Winter 2008, p. 12.

643 Todd Wilkinson. "The Maine Way." *Land & People:* Fall/Winter 2008.

644 Rachel Carley. *Wilderness A to Z.* New York: Simon & Schuster, 2001, p. 293.

645 Backpacker. "Peak Picks." *Backpacker:* February 2008, p. 63.

646 Mike Magnuson. "Retro on the Ridge." *Backpacker:* August 2007, p. 38.

647 Doug Scott. *The Enduring Wilderness.* Golden, Colorado: Fulcrum Publishing, 2004, p. 141.

"The lands of the state, now owned or hereafter acquired, constituting the forest preserve as now fixed by law, shall be forever kept as wild forest lands. They shall not be leased, sold or exchanged, or be taken away by any corporation, public or private, nor shall the timber thereon be sold, removed or destroyed."[648]

Adirondack Park, five hours north of New York City by car, encompasses 6 million acres of northern New York. A combination of public and private lands, approximately 2.5 million acres have been set aside as "forest preserve," protected under the state constitution as lands to be "forever kept as wild forest lands." The land in these areas cannot be leased, cut, dammed, or otherwise destroyed, and approximately 1 million acres of the 2.5-million-acre forest preserve have been designated as "wilderness."[649]

Interestingly, this Vermont-sized park is also the largest single protected natural area in the Lower 48 states.[650] And perhaps more important, the Adirondacks wilderness protection efforts deeply influenced future wilderness leaders like Bob Marshall and, later, Howard Zahniser, who cited this constitutional protection as an influence while they sought the strongest possible way to protect federal wilderness areas.[651]

The highest point in New York's Adirondacks is 5,344-foot Mount Marcy, which I climbed during July 2006.[652] Marcy surpasses Mount Katahdin's height by a mere 77 feet, and Katahdin, at the north end of the Appalachian Trail (AT), is only 1,485 feet taller than Springer Mountain, Georgia, at the south end. But for thru-hikers, the elevation changes in between makes hiking the AT akin to climbing Mount Everest, 14 times![653]

The AT is one of eight U.S. National Scenic Trails, which include the:[654]

- Appalachian Trail (2,175 miles)
- Pacific Crest Trail (2,650 miles)
- Florida Trail (1,400 miles)
- Ice Age Trail (1,000 miles)

---

648 Doug Scott. *The Enduring Wilderness*. Golden, Colorado: Fulcrum Publishing, 2004, p. 141.

649 George Wuerthner (ed.). *Thrillcraft*. White River Junction, Vermont: Chelsea Green Publishing Company, 2007, p. 195.

650 George Wuerthner (ed.). *Thrillcraft*. White River Junction, Vermont: Chelsea Green Publishing Company, 2007, p. 65.

651 Doug Scott. *The Enduring Wilderness*. Golden, Colorado: Fulcrum Publishing, 2004, p. 141.

652 I climbed Mount Marcy (my 42nd state highpoint) on 7/29/06.

653 Glenn Scherer. "Walking Down a Dream." *American Hiker*: Winter 2008, p. 9.

654 Glenn Scherer. "Walking Down a Dream." *American Hiker*: Winter 2008, p. 6.

- Natchez Trace Trail (500 miles)
- Potomac Heritage Trail (990 miles)
- North Country Trail (4,600 miles)
- Continental Divide Trail (3,100 miles)[655]

Bart Smith, who has hiked all eight National Scenic Trails covering over 16,400 miles, says, "To a large degree we are trending as a nation toward becoming very homogenous, with retail chains, television and such. But the National Trails take you out into the tremendous natural diversity of the country, where Americans still have their own way of talking, of acting, and where there continues to be a real diversity of people... Most Americans live within two hours of the National Scenic Trail System. So get out. Take a walk. Enjoy the natural world."[656]

In 2009, the number of northbound hikers on the Appalachian Trail spiked up to 1,425, almost 200 more than the previous year. Dave Tarasevich, a ridge runner for Baxter State Park, said most thru-hikers he's encountered over the years have been fresh out of college or retired. But in 2009, Tarasevich noticed a jump in the number of middle-aged hikers, saying, "Many of them were downsized or outsourced." On December 16, 2008, Kevin Downs (a 36-year-old laid-off engineer) made a good decision: stop looking for a job and start hiking the Appalachian Trail.[657]

"I used to be a big-house, big-car person," he said posthike. "After the trail, I'm going to live my life more simply."[658] Bravo! In addition to being the Appalachian Trail's northern terminus and a breathtakingly beautiful climb (especially in the fall), Mount Katahdin is special for another reason: it's the very first place where the sun hits the U.S. mainland each day. If that's not enough to pique your interest, it's smack-dab in the middle of Maine's vast Northwoods. I'd go back to hike, climb, or camp in Baxter State Park anytime.

In his book *Open Spaces*, Jim Dale Vickery notes that Maine's Northwoods was a place explored often by the likes of New England's favorite son, Henry David Thoreau. The author of *Walden*, *The Maine Woods*, and *A Week on the Concord and Merrimack Rivers*, Thoreau left a lasting impression in the American mind and on our public lands conservation history. During Thoreau's relatively short life (1817-1862), he developed a seminal appreciation of nature and the need for wilderness preservation that has influenced countless conservationists and naturalists, from John Muir, Aldo Leopold, Edward Abbey, and David

655 Glenn Scherer. "Walking Down a Dream." *American Hiker*: Winter 2008, p. 6.
656 Glenn Scherer. "Walking Down a Dream." *American Hiker*: Winter 2008, p. 6.
657 Backpacker magazine. "Thru-Hiking the Recession." *Backpacker*: March 2010, p. 58.
658 Backpacker magazine. "Thru-Hiking the Recession." *Backpacker*: March 2010, p. 58.

Petersen to Annie Dillard, Edward Hoagland, and Barry Lopez.[659]

Over a hundred years ago, Thoreau made a prophetic and legendary (among conservationists) statement: "In Wildness is the preservation of the world." He said this while living in a small cabin near the rural village of Concord, Massachusetts, at a time when the North American continent was comparatively unsettled and the West still largely unknown. Even then, he foresaw the need for national preserves being set aside "in which the bear and panther may still exist, and not be 'civilized off the face of the earth.'"

Shocked by his encounters with a Maine landscape badly scarred by logging, Thoreau made one of the country's first calls for wilderness preserves. In his essay *Chesunkook*, he warned against beating the rugged Maine forests into submission along the model already established in Massachusetts. "We seem to think that the earth must go through the ordeal of sheep-pasturage before it is habitable by man," he chastised.[660]

Thoreau lamented over the loss of wildlands in Maine, Massachusetts and elsewhere, and wrote, "When I consider that the nobler animals have been exterminated here... I cannot but feel as if I lived in a tamed and, as it were, emasculated country... Is it not a maimed and imperfect nature that I am conversant with?" Thoreau instinctively understood that what impacts wildlands and wildlife impacts us as well.[661]

In 1854, Thoreau published *Walden*, a classic book in which he reflects on the two years, two months, and two days he spent living near Walden Pond, explains *USA Today* contributor Mark Thoreau. Today, the 150 acres around Walden Pond are overseen by Massachusetts as a state landmark, with assistance from The Thoreau Society, a nonprofit organization that educates the public about Thoreau's works. A replica of Thoreau's simple, one-room cabin has been built for the site's many visitors, and there's hiking on the grounds throughout the year.[662]

Thoreau was a prophet of the simple life. Do your accounting, he said, on your thumbnails. He was an ecologist, activist, trailblazing environmentalist, and a defender of civil rights, publicly supporting John Brown's raid on Harpers Ferry at the onset of the Civil War.[663] He hated racial injustice as much as he loved nature. Thoreau went to jail in 1846 for refusing to pay his taxes while protesting the U.S. war against Mexico. Then, when Ralph Waldo Emerson visited him and asked, "What are you doing there?" Thoreau replied, "What are you doing out there?"

---

659 Jim Dale Vickery. *Open Spaces*. Minocqua, WI: NorthWord Press, Inc., 1991, p. 218.
660 Rachel Carley. *Wilderness A to Z*. New York: Simon & Schuster, 2001, p. 294.
661 Rodger Schlickeisen. "Our Lands, Our Wildlife." *Defenders*: Spring 2009, p. 4.
662 Mark Thoreau. "Thoreau descendant reflects on Walden Pond, Earth Day." *USA Today*: 4/20/10.
663 Jim Dale Vickery. *Open Spaces*. Minocqua, WI: NorthWord Press, Inc., 1991, p. 218.

Thoreau's standards and logic were clear and consistent: "If a man walks in the woods for half of each day, he is in danger of being regarded as a loafer, but if he spends his whole day as a speculator, shearing off those woods and making the earth bald before her time, he is esteemed as industrious and enterprising." Thoreau personified independence. When he refused to pay his taxes, Emerson offered to pay, but Henry declined.

Thoreau was persistent and stubborn, not unlike his modern-day reincarnation Edward Abbey, who explored and defended the redrock canyon country of the Southwest with Thoreau-like conviction. "At some point we must draw a line across the ground of our home and our being," Abbey wrote, "drive a spear into the land and say to the bulldozers, earthmovers, government and corporations, 'thus far and no farther.' If we do not, we shall later feel, instead of pride, the regret of Thoreau... who wrote near the end of his life, 'If I repent of anything it is likely to be my good behavior.'"[664]

"Wherever there are deer and hawks,
wherever there is liberty and danger,
wherever there is wilderness... Henry
Thoreau will find his eternal home."
—Edward Abbey

Above all else, explains *New York Times* contributor Ethan Gilsdorf, Thoreau was a seminal wilderness visionary far ahead of his time. He established a philosophical framework for wilderness preservation that preceded the Wilderness Act by over a century and required today's more studied understanding of ecosystems and biodiversity to be fully appreciated. On his deathbed (he died at age 44 of tuberculosis), Thoreau's final words were "moose" and "Indian." The land had obviously made an impression on him. "Remarkably fresh and unspoiled in many places," Gilsdorf says, "it can still make one today."[665]

In Thoreau's day, most of the U.S. was still wild and untamed, and because of this relative abundance, it was wildly underappreciated. Today, big, unpeopled tracts of forestland in the eastern United States are about as rare as open road on the New Jersey Turnpike or honesty in Washington, D.C. But in far northern Maine, roads and towns surrender to an ultrarural expanse of spruce, fir, beech, and maple heavily veined with free-flowing creeks and streams dotted with cedar swamps and beaver ponds.

Some 5 million acres in all, this region is known as the North Maine Woods, a

664 Scott Gollwitzer. "Perpetuating coal-generated power shortchanges us all." *Citizen-Times.com*: 9/2/09.
665 Ethan Gilsdorf. "Tracking Thoreau Through Maine's 'Grim and Wild' Land." *The New York Times*: 9/19/08.

broad and varied swath of unbroken forests supporting the only U.S. population of lynx east of the Mississippi River and the largest population of martens south of the Canadian border.[666] "If you look at a satellite photo of the eastern U.S., there's a big black spot," says Karen Woodsum, director of the Maine Woods Campaign. "That is the Maine Woods, an island of unbroken forest."[667]

In the 1850s, Henry David Thoreau set out to explore Maine's North Woods. But the poet-naturalist was not the first to travel the trails and tributaries of The Pine Tree State. The Wabanaki, meaning "people of the dawn," used the area's interconnected waterways for thousands of years before Thoreau gained the heights of Mount Katahdin or paddled the dark waters of the Penobscot.[668]

As the Wabanaki knew intimately, these Maine Woods—where Thoreau said one "might live and die and never hear of the United States"—cover a horizon-to-horizon and beyond stretch of wilderness similar to what I experienced growing up in northern Minnesota. I was immediately drawn to its autumn colors, glassy lakes, gurgling streams, and sweeping forest vistas. It's an attraction sometimes difficult to explain to those not raised with the outdoors at their doorstep. Many of my friends grew up in cities and suburbia and in all likelihood consider my passion for wilderness preservation and solo backcountry travel both eccentric and adventurous.

To paraphrase *Boundary Waters Journal* contributor Dick Pula, when I talk about my solo wilderness trips, some people think I'm "roughing it." But I don't go to the woods and wilderness to rough it. I do that back home in traffic jams and crowded airports while immersed in the personal and professional challenges of going to work and paying the bills. My true reason then is to "smooth it." I smooth out the rough edges of life, the ruts and wrinkles—a stone (Dick says) in a lapidary tumbler.[669]

Most people don't understand, though, and eventually blankly ask, as if there's no other way to live, "When are you going to buy a house, get married, and settle down?" As Thoreau said, "Our houses are such unwieldy property that we are often imprisoned rather than housed in them." I say, "Maybe I'll settle down when I find a woman who can be my best friend." An alien concept to many married people I suspect, as confirmed by our nation's divorce and infidelity rates.

My South African friend and Everest veteran Ronnie Muhl adds, "The problem with leading a good life, is that it usually prevents us from leading a truly

---

666 The Nature Conservancy. "Megaforest Mosaic." *Nature Conservancy*: Spring 2002, p. 24.
667 Ethan Gilsdorf. "Tracking Thoreau Through Maine's 'Grim and Wild' Land." *The New York Times*: 9/19/08.
668 Jennifer Winger. "Reviews." *Nature Conservancy*: Winter 2008, p. 77.
669 Dick Pula. "Solo Ice Travel." *The Boundary Waters Journal*: Winter 2009, p. 67.

great life."[670] Canyon country avatar Everett Ruess led a great life and wrote about his life-enhancing vagabond travels: "I enjoy beauty and the vagrant life I lead more keenly all the time. I prefer the saddle to the streetcar, and the star-sprinkled sky to the roof, the obscure and difficult trail leading into the unknown to the paved highway, and the deep peace of the wild to the discontent bred by cities." Me too.

As a teenager, I remember pondering how my life would unfold, looking at the lives of those around me for some idea, and eventually coming to the conclusion that there must be more to it than what they, for the most part, accepted as normal and satisfying. But in their defense, the meaning of life is an individual thing, and most find it in the commonplace daily routine of marriage, work, and kids. For me, it wouldn't be so easy, but whatever happened, I was determined to live life my own way. As Carl Jung said, "The shoe that fits one person pinches another; there is no universal recipe for living."

Other's recipes and expectations never resonated with me because my own goals and aspirations usually far exceeded (or were just different from) theirs, and I would gladly let them judge me critically from afar while I forged ahead on a more satisfying and adventurous path. Everett Ruess did the same: "I have always been unsatisfied with life as most people live it. Always I want to live more intensely and richly. Why muck and conceal one's true longings and loves, when by speaking of them one might find someone to understand them, and by acting on them one might discover one's self."[671]

I managed to ride out the trials, tribulations, and relationships of adolescence and early adulthood without the normal entanglements of unplanned pregnancies or marriage, occurrences that lassoed some of my friends into early, and usually unsatisfying, relationships, followed by predictably less satisfying and failed marriages. Instead, I emerged from that period as a young man who had taken his own path in life and experienced a bit of the world, one who'd had some interesting life adventures and was ready for more.

The nineteenth-century philosopher Friedrich Nietzsche knew of what I speak: "The individual has always had to struggle to keep from being overwhelmed by the tribe. If you try it, you will be lonely often, and sometimes frightened. But no price is too high to pay for the privilege of owning yourself." Jimmy Buffett reads Nietzsche too it seems: "I had always promised myself that I would not grow old like the majority of the people I see, working their asses off until their late sixties or early seventies and then retiring and going on a cruise, wondering how they let the good things in life pass them by. That was not going to be me."[672] Me neither.

670 Ronnie Muhl. *Inspiration*. Cape Town, South Africa: Inspiration at Work Publishing, 2006, p. 29.
671 Mark A. Taylor. *Sandstone Sunsets*. Salt Lake City: Gibbs-Smith Publishers, 1997, p. 65.
672 Jimmy Buffett. *A Pirate Looks at Fifty*. New York: Ballantine Publishing Group, 1998, p. 53.

Maybe I'm too much like Everett Ruess now. "I don't think I could ever settle down," Everett said. "I have known too much of the depths of life already, and I would prefer anything to an anticlimax."[673] Thank God I'm like Everett now. Scott Williamson, the first person to "yo-yo" the Pacific Crest Trail (PCT), is like Everett too. The PCT is a 2,650-mile National Scenic Trail stretching from Mexico to Canada through California, Oregon, and Washington. In writing about Williamson—who walked the entire PCT and back again (i.e., "yo-yoed" it) in less than one year (197 days), covering 5,300 miles—*Backpacker* magazine contributor Michael Darter said:

"And you think about other former girlfriends, and how women are great, but relationships are complicated, especially when you have a goal... Women are tricky. Relationships are tricky. The trail is simple... You're alone. Days and days alone. Why? Because you can't survive off the trail. Because things like steady work and marriage and a house fill you with fear, because the only place you feel safe is here, strolling through fields of golden yarrow and red maids... sleeping under wheeling constellations."[674]

Amen for men (and women) like Scott, for people who follow their dreams instead of the herd. I think I am one of them now. No compromises outside of work. No worries but my own. No remorse, for the most part. Author and humorist Lewis Grizzard once said, "Life is like a dog-sled team. If you ain't the lead dog, the scenery never changes." In my life, I'm the lead dog, and the scenery is always changing.

On October 21, I admired the scenic views from high atop Missouri Mountain, my 23rd fourteener. Missouri wasn't recognized as a separate mountain until 1956, when United States Geological Survey measurements distinguished "ridgelike" Missouri Mountain (14,067 ft.) from her northeast neighbor Mount Belford (14,197 ft.). The standard Missouri Gulch route starts out as a Class 1 stroll, but nearer the summit there are Class 2, 3, and 4 approaches.[675]

As I've already mentioned, it's possible to climb Belford, Oxford, and Missouri in a single day, but doing all three is a lengthy 14½-mile round-trip jaunt through 7,400 feet of elevation gain. One at a time (divide and conquer) is my preferred peak-bagging strategy. Missouri Mountain's Missouri Gulch route starts out by crossing over Clear Creek, then passes through a forested section and up a series of switchbacks.

673 John Nichols. *Everett Ruess*. Layton, UT: Gibbs M. Smith, Inc., 1983, p. 8.
674 Michael Darter. "The Unbearable Lightness of Being Scott Williamson." *Backpacker*: May 2005, p. 109.
675 http://www.fourteenerworld.com/.

Summitpost.org says the trail is in good condition and easy to follow as it rises above tree line into spacious Missouri Gulch. In the gulch, a sign marks the fork where one trail leads to Mount Belford and the other toward Missouri Mountain. Like Yogi Berra said, "When you get to a fork in the road, take it." The route from there is not overly difficult, but closer to the summit you'll encounter loose scree on the Class 2 Northwest Ridge. On the summit you're rewarded with views of the Three Apostles and other Sawatch Range peaks.[676]

The Three Apostles stand guard over the historic mining town of Winfield, deep in the heart of the range. Ice Mountain (13,951 ft.), its neighbor North Apostle (13,860 ft.), and West Apostle (13,568 ft.) form the forbidding Three Apostles massif. All three summits are located on Colorado's Continental Divide, and Ice Mountain easily makes the 13,800-foot cutoff for a centennial thirteener.[677]

Because I was intent on completing the fourteeners before venturing up Colorado's far more numerous thirteeners, the Three Apostles would have to wait. I started up the Missouri Mountain trail at 4:15 a.m., was standing on the summit at 8:45 a.m., started down at 9:15 a.m., and was back at the trailhead by 11:45 a.m. Next (on October 29), I climbed Quandary Peak, the only fourteener in Colorado's Tenmile Range and my last peak of the season.

Quandary is located in Summit County, southwest of the ski resort town of Breckenridge. Thanks to its broad east slope, Quandary may have the distinction of being one of the first fourteeners climbed on skis, according to the Colorado Mountain Club, and because many mining ruins dot Quandary's slopes, it was probably miners who first stood on the summit. The mountain's name came from a group of miners who found themselves in a "quandary" over the identification of a mineral on the slopes of a peak with various names—McCullough's Peak, Ute Peak, and Hoosier Peak—and Quandary stuck.[678]

When climbing relatively gentle Quandary, you might consider bringing your dog for a couple reasons. First, Quandary isn't an overly difficult climb. Second, according to Fourteenerworld.com, many people have met (or been accompanied by) "Horton," the "official" dog of this fourteener. Horton is a friendly, energetic, light colored lab-mix whose name tag identifies him as "Horton the Quandary Dog/Blue Lakes Rd."[679] Although it's considered one of Colorado's easiest fourteeners, fit for man and dog, Quandary is still dangerous, as this August 2009 *Denver Post* article ("Wyoming hiker's body recovered from Quandary peak") confirms.

---

676 http://www.summitpost.org/show/mountain_link.pl/mountain_id/342.
677 http://www.summitpost.org/show/mountain_link.pl/mountain_id/3368.
678 Colorado Mountain Club. "Quandary Peak." *Trail & Timberline*: Winter 2009, p. 16.
679 http://www.fourteenerworld.com/.

"The body of a 44-year-old Wyoming man was recovered from Quandary Peak in Summit County last night after he collapsed during his ascent of the mountain... according to Anna DeBattiste, spokeswoman for the Summit County Rescue Group. Authorities identified the hiker as Kevin Joseph Gill of Cheyenne, Wyo...."[680]

"Gill apparently collapsed during his ascent just below the summit at an elevation of about 14,000 feet, DeBattiste said... A Flight for Life helicopter from Colorado Springs flew a wilderness paramedic and three other rescuers... Another 23 rescuers from four different area mountain rescue teams responded on foot... This was the Summit County Rescue Group's seventh mission on Quandary Peak in 2009."[681]

I didn't initially plan on climbing Quandary and only drove out to the trailhead during midmorning to take a look around, but there was new snow on the ground and fresh tracks from others who had started up earlier in the morning. That, and the flawless weather, convinced me to start for the summit, even though it was nearing 10:00 a.m.

Dave Cooper (author of *Colorado Snow Climbs*) likes winter climbs best of all. "It might be cold, it might be windy, but winter can be the most rewarding time to climb Colorado's mountains," he says. "Gone are the crowds on the fourteeners. Nasty scree slopes can be hidden under a blanket of snow. Also buried are the trails, requiring (and inviting) you to make your own tracks. Our mountains never look more stunning than when adorned with a coating of the white stuff."[682]

I usually start fourteener climbs by 6:00 a.m., at the latest, which is critical for avoiding oftentimes treacherous afternoon thunderstorms. However, this time of year, the most dangerous high altitude hazard, lightning, is usually not a concern. I encountered several people coming down, but was the only one going up so late in the day. The wind near the summit looked strong, periodically kicking up swirling sheets of snow, but I pressed on and reached the top at 2:00 p.m., with some concern about a towering, fairly ominous-looking cloudbank filling a valley to the west.

If it blew over Quandary, I'd be socked-in and in a quandary. Wind-blown snow was quickly covering my tracks and the route down was neither completely obvious nor easily navigated due to the slick, snow-covered terrain. I entertained the thought of tunneling into snowdrifts near the summit if things took a turn for the worse, but wouldn't have lasted through the night. Had the weather

---

680 Kirk Mitchell. "Wyoming hiker's body recovered from Quandary peak." *The Denver Post:* 8/29/09.

681 Kirk Mitchell. "Wyoming hiker's body recovered from Quandary peak." *The Denver Post:* 8/29/09.

682 Dave Cooper (author of *Colorado Snow Climbs*). "Climbing Mountains in the Frosty Months." *Trail & Timberline:* Winter 2010, p. 14.

deteriorated, my best option might have been dropping down Quandary's steep, south-facing ridge into the basin below (assuming that's even possible), and then muddling my way out to Highway 9.

Dave Cooper adds some words of caution for those contemplating winter 14er climbs: "Don't assume that you'll be able to follow your track back down the mountain—footsteps can be filled in by the wind within minutes. Whiteout conditions can reduce visibility to zero."[683] But the clouds stayed put, the wind died down, and I reached my goal of sixteen 14ers for the season, twice as many as the year before.

It was a perfect late October day, and it doesn't get much better in Colorado or anywhere else. To quote Edward Abbey, "The mountain glittered under the sun with that harsh perfection characteristic of God's early work."[684] With the onset of winter, I wouldn't climb another fourteener for over six months. Much too long to go without the breathtaking views and vistas that make life here in Colorado so addicting, but a nice break nonetheless.

Besides, Colorado winters are harsh, as Colorado Mountain Club members Susan Paul and Lisa Heckel explain: "Colorado winters cut cold and hard as an ice axe, marking the end of the traditional hiking season. Trails are obliterated, blanketed in a thick white froth, mountain ridges draped in frosty corniced eaves, and gulleys are buried in icy slabs, becoming slick couloirs. All point upward, toward glistening peaks tipped in a shimmer of spindrift that ripples and glimmers frostily across the sapphire sky."[685]

Winters may be harsh and forbidding in the mountains of Colorado, but it's a perfect time to visit the redrock canyons of southern Utah, where I ended up spending election night car-camping in Bryce Canyon National Park wondering if Vice President Al Gore would beat his environmentally destructive opponent, George W. Bush. Although Al, at the time, was nearly as sneaky and slippery on the issues as W, he would have actually protected our public lands (nature and creation) from developers and drillers (Bush campaign contributors) instead of selling them off to the highest bidder. Based on this fact alone, he was (and is) the better man by far, as we all know now without a doubt.

It was disconcerting listening to a local politician's campaign ads on the radio, saying: "I'll keep the feds off your back." Wow! Talk about extremist wackos. And these are the people who were running for public office. In her article "Redrock Ranger," conservationist Marilyn B. Snell says southern Utah is

---

683 Dave Cooper (author of *Colorado Snow Climbs*). "Climbing Mountains in the Frosty Months." *Trail & Timberline*: Winter 2010, p. 14.

684 Edward Abbey. *The Journey Home*. New York: Penguin Books, 1991, p. 221.

685 Susan Paul and Lisa Heckel. "Avalanche Awareness First Step toward Safety in the Backcountry." *PikesPique*: December 2007/January 2008, p. 1.

red country, in the land and on the electoral map. In 2000, it turns out, voters in Kane County's striated desert of Navajo sandstone, Chinle, and Kayenta formations voted for George Bush at six times the rate of Al Gore.[686]

I guess with George Bush they got exactly what they wanted: a narrow-minded, thin-skinned, limited-intelligence zealot. And as we all know today, George went on to (as explained by *The Daily Beast*) have "a decidedly mixed record as president—losing the popular vote in 2000; barely beating a liberal aristocrat from Massachusetts in 2004; and presiding over the loss of both houses of Congress in 2006, and the White House in 2008."[687]

One of George's intellectually challenged Utah followers, state Senator Chris Buttars, went on to propose dropping 12th grade in order to alleviate his state's budget crunch and reduce the cost of public education.[688] Sounds like he may have missed out on a bit of education himself.[689] On the other hand, as Elbert Hubbard said, "Every man is a... fool for at least five minutes every day. Wisdom consists of not exceeding that limit." Senator Buttars seems to have exceeded his limit.

What these seemingly undereducated southern Utah politicians ("redrock rednecks" I heard someone call them) don't seem to understand (maybe due to skipping 12th grade) is federal lands in Utah don't belong only to them. If they did, they would be state lands. These folks apparently can't read the plain language of Utah's Statehood Act, reiterated in the state Constitution, by which Utahns renounced "all claim to unappropriated federal lands within its boundaries."[690]

Some Utahns apparently have a different (i.e., odd and perplexing) relationship to the public lands in their midst, notes *High Country News* contributor Jonathan Thompson. Namely, a fair number of them think it's theirs, sometimes for the wrecking (most notably when it comes to riding OHVs roughshod over the landscape or looting ancient cultural sites).[691] In my native state of Minnesota we have the same problem, as described by northern Minnesota writer and outdoorsman Shawn Perich.

"You don't have to walk very far down the halls of local government to find discontent with the public lands status quo. I once talked with a still-seated county commissioner who told me he believed the highest and

---

686 Marilyn B. Snell. "Redrock Ranger." *Sierra*: March/April 2004, p. 18.

687 Mat Latimer. "Karl Rove's Flameout." *The Daily Beast*: 10/22/10.

688 Jane Goetz. "Thank you, Utah, for leading the way." *High Country News*: 2/23/10.

689 The USA spent an average of $9,666 per student in elementary and secondary schools in 2006-2007. The state with the lowest per pupil spending was Utah: $5,683. See: "States with the lowest per pupil spending." *USA Today*: 12/15/09, p. 1D.

690 Darrell Knuffke. "Representative Bishop's Argument Ignores Facts About West." *Save Our Canyons*: Summer 2009, p. 14.

691 Jonathan Thompson. "The trouble with monuments." *High Country News*: 2/26/10.

best use of the Boundary Waters Canoe Area Wilderness would be to privatize it and develop the pristine lakes with homes and cabins to add to the local tax base. Never mind such a viewpoint would be whoppingly out of step with the view of most local voters, not to mention a majority of Minnesotans. You can find similar elected knuckle-draggers in most local governments."[692]

Federal lands—forests, deserts, canyons, mountains, and grasslands managed by the Bureau of Land Management, Forest Service, and National Park Service—belong to *all* Americans. Protecting Minnesota's or Utah's exceptional wilderness resources by, say, making wilderness study areas and other lands off limits to public lands-desecrating off-road vehicles and other destructive uses is not an attempt to usurp Minnesotans' or Utahns' authority over these lands, because they have no more authority than any other American.[693]

It was heartbreaking, to say the least, to be in the midst of some of the most strikingly beautiful wildlands left in the world while knowing that our (likely) next president, with the enthusiastic help of many people who lived there, would do his best to destroy them at the altars of oil and gas drillers and motorized wreck-reationists. Public lands are supposed to be multiple-use, but OHVs, oil and gas drilling, mining, and clear-cut logging preclude nearly *every other* use, so where's the multiple-use? A question we were destined to ask repeatedly during the next eight years.

I slept fitfully that election night in the snowed-covered Bryce Canyon National Park campground with only two other car-camping canyon explorers there to share in a deeply felt pain and anguish. The next day I hoped, along with the National Park Service rangers and others in the park, that whatever recount occurred would go Gore's way, but felt deep down that we were likely in for a rough and environmentally destructive four or more years. It turned out to be far worse than any of us could have imagined: a real-life nightmare that lasted an agonizingly long eight years.

> "We are, in a sense, the dumbest intelligent
> creatures ever to walk the face of this Earth."
> —Mike Adams[694]

692 Shawn Perich. "Pondering public lands as winter wanes in north country." *Outdoor News*: 3/12/10, p. 7.

693 Editorial. "Call for help: Congress member's right to request limits on OHVs." *The Salt Lake Tribune*: 11/2/07.

694 Mike Adams. "The Biofuels Scam, Food Shortages and the Coming Collapse of the Human Population." *Environmental News Network*: 4/26/08.

2000's Fourteeners (16):

The Hawaiian Islands: Oahu, Kauai, Maui, and the Big Island (19 Feb.-4 Mar.)[695]

Oahu (19 Feb.-26 Feb. & 3 Mar.-4 Mar.)

Kauai (26 Feb.-28 Feb.)

Maui (28 Feb.-1 Mar.)

Haleakala Nat. Park (2/29/00)

The Big Island (1 Mar.-3 Mar.)[696]

Hawaii Volcanoes Nat. Park (3/2/00)

Colorado: Lost Creek Wilderness Area (4/22/00)[697]

Colorado: Lost Creek Wilderness Area (4/29/00)

Point 13,762 (5/20/00)[698]

9. Mount Antero (5/20/00)[699]

---

695 My next trip to the Hawaiian Islands wouldn't be until February 2007, to visit the state highpoint.

696 The Big Island is home to Hawaii's state highpoint, Mauna Kea (13,796 ft.).

697 The 119,790-acre Lost Creek Wilderness is located in the Pike National Forest southwest of Denver.

698 Point 13,762 is a 13er north of Mount Antero.

699 I climbed Mount Antero from her north side, partly on the Cascade Trail.

Colorado: Great Sand Dunes National Monument, Rocky Mountain National Park, and Mount Evans (27 May-29 May)[700]

10. Tabeguache Peak (6/17/00)[701]

11. Blanca Peak (6/24/00)

12. Mount Shavano (7/1/00)[702]

13. Mount Princeton (7/2/00)

14. Mount Democrat (7/3/00)

15. Mount Lincoln (7/8/00)

16. Mount Bross (7/8/00)

17. Mount Yale (7/15/00)

Arizona: Kaibab Nat. Forest, the Grand Canyon (North Rim) and Cedar Breaks Nat. Monument (28 July-30 July)

18. Mount Belford (8/5/00)

19. Mount Oxford (8/5/00)

20. Mount Harvard (8/12/00)

21. Huron Peak (8/26/00)

Colorado: The Grottos (8/26/00)[703]

22. Mount Lindsey (9/3/00)

---

700 Car-camping trip with my cousin Karl Nyman.
701 Climbed Tabeguache Peak on the Southwest Ridge's Jennings Creek Trail.
702 Climbed Mount Shavano starting from the Angel of Shavano trailhead, following the Colorado Trail to the Blank Gulch trailhead, then westward and upward from there.
703 The Grottos are located east of Aspen on Highway 82. They're a series of spacious caverns cut into a granite outcropping by an earlier course of the Roaring Fork River. See Lee Gregory's Colorado Scenic Guide: Northern Region, p. 146.

Jimmy Buffett: Las Vegas (9/29/00)[704]

Maine: Baxter State Park and Mount Katahdin (7 Oct. - 9 Oct.)

Mount Katahdin (10/8/00)[705]

23. Missouri Mountain (10/21/00)

24. Quandary Peak (10/29/00)

Utah/Arizona: San Rafael Swell (Buckhorn Draw and Little Wild Horse Canyon), Capitol Reef Nat. Park, Grand Staircase Escalante Nat. Mon. (Bull Valley Gorge and Grosvenor Arch), Bryce Canyon Nat. Park, Page, Ariz., area and the Glen Canyon Dam, Navajo Nat. Mon., and Canyon de Chelly (pronounced de-SHAY) Nat. Mon. (4 Nov.-10 Nov.)[706]

(To be continued in 4/44/14 II: Nemesis)

---

704 My first Jimmy Buffett concert: At the MGM in Las Vegas.
705 I completed this Mount Katahdin climb in 11 hours on the nine-mile-plus (round-trip) Knife Edge route, hiking-climbing through fresh snow the entire day. Mount Katahdin was my 3rd state highpoint.
706 I spent election night car-camping in Bryce Canyon National Park, Utah.

"To dare! To dare! Ever to dare! said Danton, speaking of
revolution and giving us the metaphor we need. Reproduction
and mere survival never have been good enough for humankind.
Even a simple hike... involves that element of risk and effort which
compensates for the usual banality of our lives. *We love the taste
of freedom. We enjoy the smell of danger.* We take pleasure in
the consummation of mental, spiritual, and physical effort; it is the
achievement of the summit that brings the three together."
–Edward Abbey, *The Journey Home*[707]

"Accidents in the mountains are less common than in the lowlands,
and these mountain mansions are decent, delightful, even divine,
places to die in, compared with the doleful chambers of civilization.
Few places in this world are more dangerous than home. Fear
not, therefore, to try the mountain-passes. They will kill care,
save you from deadly apathy, set you free, and call forth every
faculty into vigorous, enthusiastic action."
–John Muir, *The Wild Muir*[708]

---

707 Edward Abbey. *The Journey Home*. New York: Penguin Books, 1991, p. 215.
708 Lee Stetson (ed.). *The Wild Muir*. Yosemite National Park, CA: Yosemite Association, 1994, p. ix.

"I climbed higher... The sun was just rising as I reached
the summit. Every muscle in my body thrilled to the wild
and mysterious feeling that comes only to one who can
get into the wildest places, farthest from other people."
–Kent Frost, *My Canyonlands*[709]

"The love of wilderness is more than a hunger for what is always
beyond reach; it is also an expression of loyalty to the earth, the earth
which bore us and sustains us, the only paradise we shall ever know,
the only paradise we ever need–if only we had eyes to see... No,
wilderness is not a luxury but a necessity of the human spirit, as vital to
our lives as water and good bread."
–Edward Abbey, *Desert Solitaire*

---

709 Kent Frost. *My Canyonlands*. New York: Abelard-Schuman, 1971, p. 43.

David on the summit of Mount Lindsey with the Spanish Peaks in background: 3 Sep. 2000

Index

Y

Z

LaVergne, TN USA
11 January 2011
211969LV00003B/9/P